Superfast Physics for 14 to 16 year olds

By Michael D. Reid

Catchy memory tricks
and hundreds of practice questions

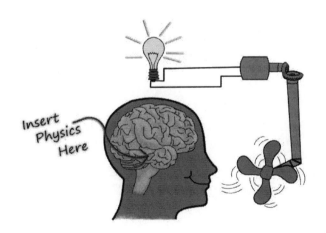

Superfast Physics for 14 to 16 year olds

Written by Michael D. Reid
Edited by Alysia Relf-Phypers

Content copyright © 2018 Michael D. Reid.
All rights reserved

Use of this book for classroom purposes is encouraged with a valid purchase.

First Published November 2018

Revised and improved from 'The Ultimate Guide to Remembering GCSE Physics' © 2018 and 'The Ultimate Guide to Remembering High School Physics Fast' © 2017 Michael D. Reid

All memory aids and images are property of Michael D. Reid.

Special Offers

This book has been turned into an online course which is available on Udemy.com. It is also called, 'Superfast Physics for 14 to 16 year olds'.

The course includes multiple choice quizzes, video explanations and discussion boards to ask questions.
- Use the coupon code: SUPERFAST_BOOK to receive the course for 95% off its regular price!

Visit our website and contact us at www.123tutor.me.
- Mention this book for an extra 10% off any of our advertised tutoring packages or use the coupon code: SUPERFAST_BOOK
- One coupon per student.

Contents

Introduction 7

1. Speed, Distance, Force and Weight
Where's My Guitar? 8
The Triangle Method of Solving Problems 10
Moving in a Circle 11
Newton's Laws 12
How Far Has It Gone? 14
The Powers of 10 and Alien Prefixes 15
Approaching Long Answer Questions 17

Examples: Speed, Distance, Force and Weight 19
Questions: Speed, Distance, Force and Weight 25
Answers: Speed, Distance, Force and Weight 28
Bonus Questions 1.1: Speed, Distance, Force and Weight 37
Bonus Questions 1.2: Speed, Distance, Force and Weight 41

2. Equations of Motion and More Complex Movement
Ugly and Amazing Teapots 45
The Square Method for Tricky Equations 47

Examples: Equations of Motion and More Complex Movement 50
Questions: Equations of Motion and More Complex Movement 59
Answers: Equations of Motion and More Complex Movement 61
Bonus Questions 2.1: Equations of Motion and More Complex Movement 68
Bonus Questions 2.2: Equations of Motion and More Complex Movement 70

3. Momentum, Force, Moment and Mass
Vikings Like Dancing 72
Your Favourite Food 73

Examples: Momentum, Force, Moment and Mass 74
Questions: Momentum, Force, Moment and Mass 77
Answers: Momentum, Force, Moment and Mass 79
Bonus Questions 3.1: Momentum, Force, Moment and Mass 87
Bonus Questions 3.2: Momentum, Force, Moment and Mass 89

4. Kinetic Energy, Potential Energy and Energy Stored in Mass
The Animals' Lunchtime 91
Colin Has Energy 92

Examples: Kinetic Energy, Potential Energy and Energy Stored in Mass 95
Questions: Kinetic Energy, Potential Energy and Energy Stored in Mass 97
Answers: Kinetic Energy, Potential Energy and Energy Stored in Mass 98
Bonus Questions 4.1: Kinetic Energy, Potential Energy and Energy Stored in Mass 102
Bonus Questions 4.2: Kinetic Energy, Potential Energy and Energy Stored in Mass 104

5. Pressure in Solids, Liquids and Gas
Dancing and Pivoting 106

Examples: Pressure in Solids, Liquids and Gas 107
Questions: Pressure in Solids, Liquids and Gas 110
Answers: Pressure in Solids, Liquids and Gas 111
Bonus Questions 5.1: Pressure in Solids, Liquids and Gas 116
Bonus Questions 5.2: Pressure in Solids, Liquids and Gas 118

6. Work, Energy, Power and Heat
Colin Changes Tune 120

Examples: Work, Energy, Power and Heat 121
Questions: Work, Energy, Power and Heat 125
Answers: Work, Energy, Power and Heat 126
Bonus Questions 6.1: Work, Energy, Power and Heat 130
Bonus Questions 6.2: Work, Energy, Power and Heat 132

7. Radioactivity
The Periodic Table Game 134
Writing the Element 139
The Radiation Story 140
How is it Stopped? 142
Radioactive Decay Equations 143
Subatomic Particles 145

Examples: Radioactivity 148
Questions: Radioactivity 155
Answers: Radioactivity 156
Bonus Questions 7.1: Radioactivity 162
Bonus Questions 7.2: Radioactivity 164

8. Charge, Energy, Efficiency and Electrical Power 1
Rabbits and Carrots 166
Electric Toothpaste 168

Examples: Charge, Energy, Efficiency and Electrical Power 1 170
Questions: Charge, Energy, Efficiency and Electrical Power 1 175
Answers: Charge, Energy, Efficiency and Electrical Power 1 176
Bonus Questions 8.1: Charge, Energy, Efficiency and Electrical Power 1 180
Bonus Questions 8.2: Charge, Energy, Efficiency and Electrical Power 1 182

9. Resistors in Series and Parallel, Voltage Across Resistors
Sapud the Wonderfish 184
Rusty Robots 185

Examples: Resistors in Series and Parallel, Voltage Across Resistors 186
Questions: Resistors in Series and Parallel, Voltage Across Resistors 192
Answers: Resistors in Series and Parallel, Voltage Across Resistors 193

  Bonus Questions 9.1: Resistors in Series and Parallel, Voltage Across Resistors 199
  Bonus Questions 9.2: Resistors in Series and Parallel, Voltage Across Resistors 201

10. Electrical Power 2 and Transformers
  The Power of Imagination 203
  Equations for Transformers: Vampires and Vans 205

  Examples: Electrical Power 2 and Transformers 210
  Questions: Electrical Power 2 and Transformers 215
  Answers: Electrical Power 2 and Transformers 216
  Bonus Questions 10.1: Electrical Power 2 and Transformers 221
  Bonus Questions 10.2: Electrical Power 2 and Transformers 223

11. Waves, Light and Colours
  New Rules 225
  Short Gavin 227
  Which Path? 228
  Orange Rabbits 232

  Examples: Waves, Light and Colours 233
  Questions: Waves, Light and Colours 238
  Answers: Waves, Light and Colours 240
  Bonus Questions 11.1: Waves, Light and Colours 248
  Bonus Questions 11.2: Waves, Light and Colours 251

12. More Complicated Systems
  Nutcrackers as Force Multipliers 255
  A Hydraulic Force Multiplier 256
  Wire Travelling in a Magnetic Field 258
  An Electric Generator 259
  The Magnetic Field of a Current in a Wire 260
  An Electric Motor 261
  Electromagnets 262
  A Transformer 263

  Questions: More Complicated Systems 264
  Answers: More Complicated Systems 265

13. Bonus Material
  One Memory Aid to Rule Them All 266
  A Final Note 274
  All the Equations in One Place 275
  Calculating Area and Volume 277
  Answers to Bonus Questions 278

Introduction

Physics is easy if you have the right tools and you practice what you know. Each chapter of this book has given you the tools to learn the equations or key points in the topic covered. Every section starts with at least one memory aid for remembering the key information. Repeat the phrases and write down their meanings so that you are comfortable. Once you know the memory aids, you can then progress to the worked through examples. These are the tools you need to succeed in physics at this level.

You practice using these tools by working through the questions that follow and checking your answers with the fully worked answers that come afterwards. Once you are confident that you can answer all of the questions, you can take the training wheels off and work on the bonus questions. The answers for these are in the back of the book.

Practice is the key.

The first 11 chapters of this book are based around the main topics of physics for 14 to 19 year old students. Chapter 12 covers more complicated systems in as simple a manner as possible and chapter 13 is bonus material which includes a story to help you remember all of the smaller memory aids.

This book is designed to help you in a way which textbooks are unable. It is a book about memory techniques, physics equations and empowerment through confidence and practice. What this book does is it utilizes our amazing minds to create new connections. You are more likely to be able to remember something if it is fun and if you are interested in it. By using these memory aids, you are creating new connections in your brain that make learning the physics equations easy.

Let's start and not waste time.

1. Speed, Distance, Force and Weight

Where's My Guitar?

<u>S</u>ing <u>a</u> <u>t</u>une
<u>d</u>own <u>t</u>he <u>s</u>treet.
<u>F</u>ind <u>a</u> <u>m</u>an
<u>w</u>ith <u>m</u>y guitar.

For this section, each of the underlined letters stands for something in an equation. This can be written in a triangle form which makes it easy to use. This is explained after we look at what the little memory aid actually means.

<u>S</u>ing <u>a</u> <u>t</u>une
<u>d</u>own <u>t</u>he <u>s</u>treet.
<u>F</u>ind <u>a</u> <u>m</u>an
<u>w</u>ith <u>m</u>y <u>g</u>uitar.

<u>S</u>ing*	s	Speed / Acceleration × Time
<u>a</u> <u>t</u>une	a × t	
<u>d</u>own	d	Distance / Time × Speed
<u>t</u>he <u>s</u>treet.	t × s	
<u>F</u>ind	f	Force / Acceleration × Mass
<u>a</u> <u>m</u>an	a × m	
<u>w</u>ith	w	Weight / Mass × Gravity
<u>m</u>y <u>g</u>uitar.	m × g	

This can be used for either speed (a scalar quantity) or velocity (a vector quantity). Remember that velocity is just speed with a direction.

This is the first memory aid. Practice it and practice writing down the full equations in words as well. Practice it and repeat it until you know each of the equations by heart. This is important because in later memory aids a letter may be reused with a different meaning. Practice is the key here.

The Triangle Method of Solving Problems

If I put the equation that I am using into a triangle (here the equation is (change in) speed = acceleration x (change in) time), then I can use a triangle to find the information I want.

If I want to find acceleration from the equation, then I just cover up acceleration with my finger. The section that remains shows me the calculation that I need to perform.

This is telling me that:

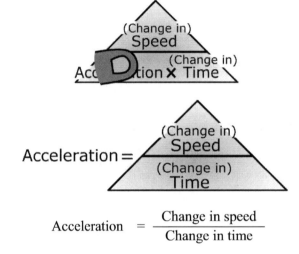

$$\text{Acceleration} = \frac{\text{Change in speed}}{\text{Change in time}}$$

If I want to find the time taken (change in time) then I just cover up the change in time with my finger.

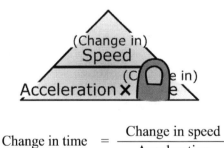

$$\text{Change in time} = \frac{\text{Change in speed}}{\text{Acceleration}}$$

If I want to find the change in speed then again, I cover up the part I want to find (change in speed) and what's left gives me the equation I use to get it.

Change in speed = Acceleration x Change in time

Moving in a Circle

When an object moves in a circle anti-clockwise it moves up, then left, then down, then right and then back to the start. The speed remains the same during this time but the velocity is changing because the direction of the velocity is always changing (i.e. up, left, down, right).

When an object is moving in a circle the acceleration is always towards the centre of the circle. This means that the force experienced is also always towards the centre of the circle. Like velocity, the direction of the acceleration and force is always changing even though the magnitude remains the same.

All objects moving in a circle have a force and acceleration towards the centre of the circle.

Remember your <u>fav</u>ourite <u>speed</u>

Vectors: - **F**orce   Scalar: - **Speed**
                 - **A**cceleration
                 - **V**elocity

Example:

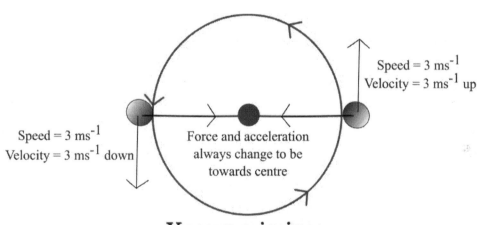

- Speed = 3 ms^{-1}, Velocity = 3 ms^{-1} down
- Force and acceleration always change to be towards centre
- Speed = 3 ms^{-1}, Velocity = 3 ms^{-1} up

You are swinging a ball over your head

Newton's Laws

Also here are a couple of ideas that are important for understanding this topic.

- Some of the equations in this topic are based off of the famous Newton's laws. You need to know these!

- It's not normal for things to sink or fall through tables, or the ground, or rocks... The force that stops this is called the 'normal force'. On a flat surface it is the same amount as the weight but in the opposite direction.
    - This is from Newton's first law and Newton's third law.

- Any unbalanced force will give an acceleration.
    - This is from Newton's second Law.

- Most of the time we can take the value of gravity to be 10 ms^{-2}. It is a type of acceleration.

Newton's 1st law: An object at rest will remain at rest and an object in motion will continue with the same motion as long as the forces are balanced.

Newton's 2nd law: If an unbalanced force acts on an object, that object will accelerate in the same direction as the force according to $F = ma$.

Newton's 3rd law: For every action force there is an equal and opposite reaction force.

So how do you remember these? There is a famous story of Newton discovering gravity after he got hit on the head by a falling apple – this will help us.

Stage 1 – Rest

Newton is under the tree but the apple isn't moving: The apple remains at rest.

Newton's 1st law: An object at rest will remain at rest and an object in motion will continue with the same motion as long as the forces are balanced.

Stage 2 – Acceleration

Newton remains under the tree but a breeze causes the apple to fall. It accelerates towards his head.

Newton's 2nd law: If an unbalanced force acts on an object, that object will accelerate in the same direction as the force according to $F = ma$.

Stage 3 – Collision

Newton gets hit on his head by the apple and feels the force. At the same time the apple experiences the same force but in the opposite direction.

Newton's 3rd law: For every action force there is an equal and opposite reaction force.

How Far Has It Gone?

If you want to find the **distance travelled**, you have to **underst**and it.

Find the **distance travelled** by looking at the area **under** the **s**peed-**t**ime graph.

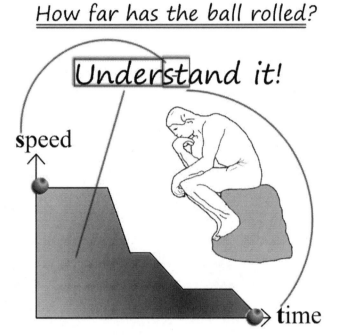

Another way to find distance travelled, especially when examining a speed-time or velocity-time graph is just to find the area under the speed-time or velocity-time graph between the times you are interested in. We will show this in the example questions for this first section.

The Powers of 10 and Alien Prefixes

The powers of 10 are also important in physics. For example, we measure distance in metres. However large distances may be measured in kilometres. This means 'thousand metres'.

The different prefixes (the kilo part) go up and down by a thousand each time. 1 kilobyte is 1,000 bytes. 1 megabyte is 1,000 kilobytes. 1 gigabyte is 1,000 megabytes. We also like to use centi as in centimetres. This is an exception as it's a hundredth while milli is a thousandth.

Another memory aid:

<u>9</u> <u>g</u>reen <u>m</u>en <u>k</u>now <u>no</u>thing ex<u>ce</u>pt <u>m</u>aybe <u>m</u>y <u>n</u>ame

(9) Green **m**en **k**now **no**thing ex**ce**pt **m**aybe **m**y **n**ame

10^9 **G**iga, 10^6 **M**ega, 10^3 **k**ilo, **No** prefix, 10^{-2} **ce**nti, 10^{-3} **m**illi, 10^{-6} **m**icro, 10^{-9} **N**ano

Prefix	Symbol	Quantity	Full Quantity
Giga	G	10^9	1 000 000 000
Mega	M	10^6	1 000 000
kilo	k	10^3	1 000
no prefix, just the amount	No symbol	10^0	1
centi	c	10^{-2}	0.01
milli	m	10^{-3}	0.001
micro	µ	10^{-6}	0.000 001
nano	n	10^{-9}	0.000 000 001

Approaching Long Answer Questions

Long answer questions cause a lot of problems for people. Often the first question students have is what do they have to write to get the full set of marks available.

Students will then often write answers which can be a page long for only a few marks when some well-structured sentences would do equally as well or better.

The following is a question that students of all abilities from schools all over the world struggle with. The question itself is simple and it is only worth 5 marks.

What I want you to do is to attempt to write the answer before you read the technique that will allow you to answer it reliably.

Question: Explain what happens when a glass is pushed off of a table. (5 marks)

Don't worry if you only get 1, 2 or even no marks for your answer. This is normal.

I use this question because EVERYBODY knows EXACTLY what happens when a glass is pushed from a table. What we are really interested in is finding out how we can get the 5 marks. Try and write the answer to this question now. Set a time limit of between 5-10 minutes, it can be done in 30 seconds but if you are competitive you will not want to stop until you get the full answer!

Try this now.

Here is the technique for answering any longer questions. The longer answers should always be structured in the following way.

It is our job to identify what has happened, the effects of this, what they mean and the result of these changes:

    1. Something happens
    2. Which has an effect
    3. Which means
    4. Therefore

Let's look at how this works. Mark your answer as you work through it. Be honest and don't worry if you didn't score a good mark. That's why we're running through this.

Question: Explain what happens when a glass is pushed off of a table. (5 marks)

1. *Something happens:*
The glass has been pushed off the table. (No marks – this is just repeating part of the question)

2. *Which has an effect:*
The glass' weight (the force due to gravity) is no longer balanced by the normal reaction force from the table. (1 mark)

3. *Which means:*
The glass has an unbalanced force acting on it (its weight). (1 mark)

4. *Therefore:*
The glass will accelerate towards the ground (1 mark) in the direction of its weight. (1 mark) The glass may break when it hits the ground (1 mark – this mark is only gained if it is stated that the cup may break, not will break)

So the final answer would be written as:

Question: Explain what happens when a glass is pushed off of a table. (5 marks)

Answer:
The glass has been pushed off the table. The glass' weight is no longer balanced by the normal reaction force from the table. The glass has an unbalanced force acting on it (its weight). The glass will accelerate towards the ground in the direction of its weight. The glass may break when it hits the ground.

Examples: Speed, Distance, Force and Weight

Example 1) A runner has travelled 3 kilometres. It has taken 15 minutes. What was the speed of the runner?

speed = $\dfrac{\text{distance}}{\text{time}}$

    time (in seconds) = 15 minutes x 60 seconds per minute
                         = 900 seconds

    distance (in metres) = 3 km x 1,000 km per metre
                         = 3,000 m

speed = $\dfrac{3{,}000 \text{ m}}{900 \text{ s}}$

     = 3.33 ms^{-1}

This question is interesting in that the distance travelled is given in kilometres and the time is given in minutes. This means that, if we wanted to we could have also calculated an answer in km/h.

For this we need time as a fraction of an hour.

time (in hours) = 15 minutes

               = $\dfrac{15 \text{ min}}{60 \text{ min/h}}$

               = 0.25 h

speed = $\dfrac{3 \text{ km}}{0.25 \text{ h}}$

     = 12 km/h

Both of these answers are equally valid.

Example 2) A section of plastic pipe is accelerating downhill away from a construction worker. It accelerates from rest at a constant rate of 1 ms^{-2}. It takes the pipe 8 seconds to reach the bottom of the hill. What is its final speed?

speed = acceleration x time
      = 1 ms^{-2} x 8 s
      = 8 ms^{-1}

Example 3) A child is sitting on a skateboard being pulled by his brother. He has a mass of 20 kg and he is accelerating at 3 ms^{-2}.

a) What force is being provided by his brother?

force = acceleration x mass
     = 3 ms^{-2} x 20 kg
     = 60 N

b) His brother is becoming tired and can now only pull with a force of 10 newtons. What is the acceleration of the child on the skateboard?

$$\text{acceleration} = \frac{\text{force}}{\text{mass}}$$

$$= \frac{10 \text{ N}}{20 \text{ kg}}$$

$$= 0.5 \text{ ms}^{-2}$$

Example 4) A car has a mass of 1,400 kg. What is the weight of the car? Take gravity on Earth to be 10 ms^{-2}. (You will need to remember this value for gravity.)

weight = mass x gravity
= 1,400 kg x 10 ms^{-2}
= 14,000 N

Example 5) A train accelerates from the station, where it had been stopped, at a constant rate of 0.75 ms^{-2} until it reaches a speed of 27 ms^{-1}. How long did it accelerate for?

$$\text{time} = \frac{\text{speed}}{\text{acceleration}}$$

$$= \frac{27 \text{ ms}^{-1}}{0.75 \text{ ms}^{-2}}$$

$$= 36 \text{ s}$$

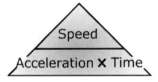

Example 6) A ball is accelerated at a rate of 30 ms^{-2} by a force of 4 newtons. Calculate the mass of the ball.

$$\text{mass} = \frac{\text{force}}{\text{acceleration}}$$

$$= \frac{4 \text{ N}}{30 \text{ ms}^{-2}}$$

$$= 0.13 \text{ kg}$$

Example 7) A car accelerates from 10 ms^{-1} to 20 ms^{-1} in a time of 4 seconds. What is the value of the car's acceleration?

$$\text{acceleration} = \frac{\text{(change in) speed}}{\text{time}}$$

$$= \frac{20 \text{ ms}^{-2} - 10 \text{ ms}^{-2}}{4 \text{ s}}$$

$$= 2.5 \text{ ms}^{-2}$$

Example 8) A truck accelerates to a speed of 15 ms^{-1} in a time of 15 seconds. It then travels on at a constant speed. Using the speed time graph calculate the following.
a) The distance travelled in the first 15 seconds.
b) The distance travelled in the first 30 seconds.
c) The average acceleration of the truck between 5 and 10 seconds.

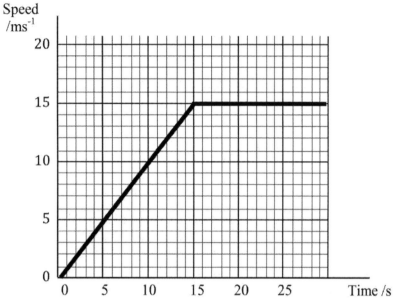

a) We are interested in finding the area under the graph during the first 15 seconds. As this is a triangular region we just calculate the area of the triangle.

area of a triangle = ½ x base x height
= ½ x 15 s x 15 ms^{-1}
= 112.5 m

This is the distance travelled. However, normally we only use 3 significant figures in our answer. Thus the distance travelled is then:

distance = 113 m

b) Again we need to find the area under the graph during the 30 seconds. We already have the area for the first 15 seconds so we need to only calculate the area under the graph between 15 s and 30 s and add it to the answer in part a.

area under graph between 15 s and 30 s = (30 s – 15 s) x 15 ms^{-1}
= 225 m

total area = area between 0 s and 15 s + area between 15 s and 30 s
= 113 m + 225 m
= 338 m

total distance travelled in the first 30 seconds = 338 m

c) We are interested in the acceleration in the region of the graph between 5 s and 10 s.

$$\text{acceleration} = \frac{\text{(change in) speed}}{\text{(change in) time}}$$
$$= \frac{10 \text{ ms}^{-1} - 5 \text{ ms}^{-1}}{10 \text{ s} - 5 \text{ s}}$$
$$= 1 \text{ ms}^{-2}$$

Example 9) A jug has water added to it. Initially its weight is 9.2 N. Once the water is added the weight of the jug becomes 15.3 N. What mass of water was added to the jug?

weight of water added to jug = weight of the jug + water − weight of the jug
$$= 15.3 \text{ N} - 9.2 \text{ N}$$
$$= 6.1 \text{ N}$$

This gives us the weight of the water but we need its mass.

$$\text{mass of water} = \frac{\text{weight}}{\text{gravity}}$$
$$= \frac{6.1 \text{ N}}{10 \text{ ms}^{-2}}$$
$$= 0.61 \text{ kg}$$

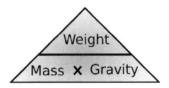

Example 10) A bicycle is travelling at 15 ms^{-1}. The cyclist sees a small animal 45 m in front of him and pulls the brakes. His reaction time from seeing the animal to applying his brakes is 0.3 seconds. He slows down at a constant rate of 3 ms^{-2}. Will he stop in time to miss the animal?

* A note on the above question. Although the final answer will be a yes or a no, without the full set of calculations you will not get the mark for the final answer. This is because with no calculations it is just a guess with none of the hard work. A guess never built a bridge or fixed a car, but hard work on the other hand...

First here we need to sketch the speed time graph. We are interested in the distance travelled so this is given by the area under the graph.

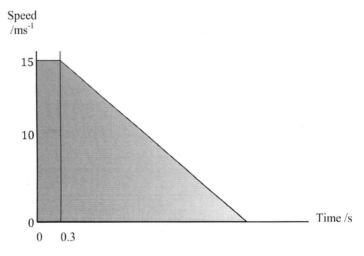

We can see that the graph is divided nicely into 2 sections: the time when the cyclist is reacting to seeing the animal and the time where the cyclist is slowing down after pressing the brakes.

The first area can be calculated immediately
area = 0.3 s x 15 ms^{-1}
= 4.5 m

The second area requires a little more information. We need to find out how long it takes the cyclist to change his speed by 15 ms^{-1} when they are accelerating at 3 ms^{-2}.

time = $\frac{\text{(change in) speed}}{\text{acceleration}}$

= $\frac{15 \text{ ms}^{-1}}{3 \text{ ms}^{-2}}$

= 5 s

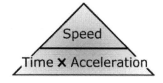

We now have enough information to find the second area under the graph.

area of triangle = ½ x 15 ms^{-1} x 5 s
= 37.5 m

So the total distance travelled = 4.5 m + 37.5 m
= 42 m

We can now see that as the animal was 45 m away and the bike needed 42 m to stop, the small animal will survive.

Example 11) Which of the following is true for motion in a circle?
   a) The acceleration is towards the centre and the velocity is constant.
   b) The force is away from the centre and the velocity is changing.
   c) The acceleration is away from the centre and the speed is constant.
   d) The force is towards the centre and the velocity is changing.

For objects moving in a circular motion, only speed is a scalar quantity. This means that it has only a magnitude and no direction. Everything else is a vector quantity and changes direction as the object moves because the direction is always towards the centre.

The answer is D.

Use the graph below to answer the next 3 questions.

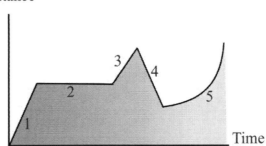

A hamster has been let out to play by its owner. The above graph shows its motion when it was running about the room. The distance is measured from its owner who is standing still.

Example 12) Which parts of the above graph refer to a period where the hamster must be moving?
   a) 1,2,4,5
   b) 1,2,3,4,5
   c) 1,3,4,5
   d) 5
   e) None of the above

This graph shows the change in distance from a fixed point over time. In order to show movement, the line has to have any gradient which is not 0. This means that the distance is changing.
- The answer is C.

Example 13) Which parts of the above graph refer to a period where the hamster must be accelerating?
   a) 1,2,4,5
   b) 1,2,3,4,5
   c) 1,3,4,5
   d) 5
   e) None of the above

If you divide distance by time you get speed, this is the gradient. The sections of the graph that have a straight-line gradient show constant speed. This means that there is no acceleration. When something accelerates the speed also changes This means that the gradient changes and so the line would curve. This only happens at section 5.
- The answer is D.

Example 14) Which parts of the above graph refer to a period where the hamster could be stationary?
   a) 1,3
   b) 1,2,3,4,5
   c) 1,3,4,5
   d) 2
   e) None of the above

The y-axis is labelled 'distance' which means that it is showing how far the hamster has moved from a fixed position. The only time that shows no movement is when the line is flat. That is section 2.
- The answer is D.

Questions: Speed, Distance, Force and Weight

Now let's look at some more questions. The fully worked answers are after the questions. Try to do as many as you can before looking there.

1.1) A motorbike accelerates from rest to a speed of 20 ms^{-1} in 5 seconds. It continues at this speed for 8 seconds and then slows to a stop in a further 7 seconds. The total trip takes 20 seconds. Calculate:
   a) The initial acceleration.
   b) The acceleration during the final part of the journey.
   c) The total distance travelled.

1.2) A car is driving along a street accelerating. It passes lamp posts that are 30 metres apart. It starts at a lamp post and travels past another 10 lamp posts in a time of 20 seconds. Calculate the average speed of the car.

Which of the following statements is correct?
   a) The time taken to travel between lamp post 1 and 2 is the same as the time taken to travel between lamp post 2 and 3.
   b) The time taken to travel between lamp post 1 and 2 is greater than the time taken to travel between lamp post 2 and 3.
   c) The time taken to travel between lamp post 2 and 3 is greater than the time taken to travel between lamp post 1 and 2.
   d) The distance between the lamp posts is changing.

1.3) A go cart is travelling down a hill with constant acceleration.
At time t = 3 seconds it has a velocity of 12 ms^{-1}. At time t = 6 seconds it has a velocity of 24 ms^{-1}.
   a) Calculate the acceleration of the go cart.
   b) Calculate the distance travelled during this time.

1.4) A motorbike is accelerating from rest. It reaches a velocity of 35 ms^{-1} after 12 seconds and then continues at this speed for 20 seconds before slowing to a stop 40 seconds after it began to move.
   a) Calculate the initial acceleration.
   b) Calculate the final acceleration.
   c) Calculate the total distance travelled.

1.5) A sports fan is viewing a race through some binoculars. She sees the flash from the starting gun and then after 4 seconds she hears the sound. The speed of sound is 340 ms^{-1}.
   a) Complete the table.

Speed of sound /ms^{-1}	Distance to travel /m	Time for sound to travel the distance /s
340	340	1
340	680	
340		3
340		4

   b) How far away is the sports fan from the racers?
   c) What assumptions have been made in this calculation?

1.6) An empty bus accelerates from a bus stop and travels along a street to the next bus stop. There it picks up 20 people and accelerates from the bus stop. The engine is producing the same amount of force to move the bus however the bus accelerates more slowly. Why?

1.7) The acceleration of the moon is 1.62 ms^{-2}. A golfer from Earth has travelled to the moon to take part in a golf tournament. On Earth he is normally able to hit a golf ball 240 yards. He hits the ball on the moon and is surprised at how far the ball travels. Explain this.

(Disclaimer: there are no golf courses on the moon! This is just an interpretive question about why this would occur. Also this question is about interpreting and explaining change in distance. This means that there is no need to use metres, although they would be equally valid. Any distance type could be used here (although leagues, perches and parsecs would be unlikely to be used in measuring golf drive distances)).

1.8) A car is broken down. A pedestrian offers to help. The driver believes that if the car can be pushed to a speed of 2 ms^{-1} he will be able to start it. The pedestrian agrees to push the car but asks the driver to remove all of his heavy shopping first. Why is this a good idea?

1.9) A girl has a mass of 40 kg. Calculate her weight on the following planets.
        Pluto, gravity = 0.9 ms^{-2} Earth, gravity = 10 ms^{-2}
        Venus, gravity = 9 ms^{-2} Jupiter, gravity = 25 ms^{-2}

1.10) The engine on a moving car develops a force of 2,000 N accelerating it forward. When the car has a speed of 30 ms^{-1} it no longer accelerates.
    a) Calculate the resistance force acting on the car.
    b) What happens in the following circumstances?
        i) The resistive force acting on the car decreases.
        ii) The resistive force acting on the car increases.
        iii) The resistive force acting on the car remains the same.

1.11) Which of the following is true for motion in a circle?
    a) The speed is changing and the velocity is constant.
    b) The speed is constant and the velocity is changing.
    c) The acceleration is constant and the force is changing.
    d) The acceleration is changing and the force is constant.

1.12) The following diagram shows a toy that is moving clockwise in a circle. Which of the following options is accurate?

    a) 1 is velocity, 2 is speed
    a) 2 is velocity, 3 is acceleration
    b) 3 is force, 4 is velocity
    c) 4 is velocity, 1 is speed

Use the graph below to answer the next 4 questions.

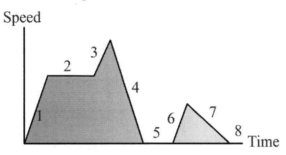

The above graph shows the motion of a cat playing with a toy. Distance is measured from a point in the centre of the room.

1.13) Which list identifies all of the areas where the cat was definitely accelerating?
 a) 1,3,4,6,7
 b) 2,5,8
 c) 1,3,6
 d) 4,7
 e) None of the above

1.14) Which list identifies the times when the cat was at its original location?
 a) 2,5,8
 b) 1,3,6
 c) 5,8
 d) 4,7
 e) None of the above

1.15) At one point the cat is running in a circle at constant rate about the measurement point. When was this?
 a) 1
 b) 2
 c) 5
 d) 8
 e) None of the above

1.16) When was the deceleration of the cat the greatest?
 a) 1
 b) 3
 c) 4
 d) 7
 e) None of the above

Answers: Speed, Distance, Force and Weight

1.1) A motorbike accelerates from rest to a speed of 20 ms^{-1} in 5 seconds. It continues at this speed for 8 seconds and then slows to a stop in a further 7 seconds. The total trip takes 20 seconds. Calculate:
   a) The initial acceleration.
   b) The acceleration during the final part of the journey.
   c) The total distance travelled.

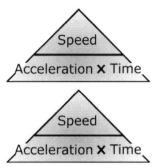

a) acceleration = change in speed /change in time
   = (20 ms^{-1}) / 5 s
   = 4 ms^{-2}

b) acceleration = change in speed / time
   = (final speed – initial speed) / time
   = (0 ms^{-1} – 20 ms^{-1}) / 7 s
   = -2.8 ms^{-2}

c) First step, draw the speed time graph. Then calculate the area under the speed time graph.

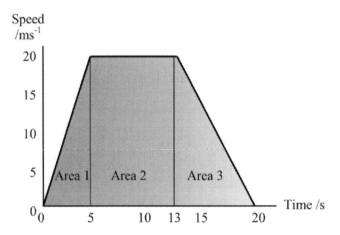

area under the graph = area 1 + area 2 + area 3
area 1 = ½ x 5 s x 20 ms^{-1}
(area of a triangle = ½ base x height)
area 1 = 50 m

area 2 = 8 s x 20 ms^{-1}
(area of rectangle = base x height)
area 2 = 160 m

area 3 = ½ x 7 s x 20 ms^{-1}
(area of triangle = ½ base x height)
area 3 = 70 m

area under graph = 50 m + 160 m + 70 m
                 = 280 m
(area under graph = distance travelled)

The total distance travelled in the journey is 280 m

1.2) A car is driving along a street accelerating. It passes lamp posts that are 30 metres apart.

It starts at a lamp post and travels past another 10 lamp posts in a time of 20 seconds. Calculate the average speed of the car.

$$\text{speed} = \frac{\text{distance}}{\text{time}}$$

distance = 10 lamp posts x 30 metres apart
= 300 m

$$\text{speed} = \frac{300 \text{ m}}{20 \text{ s}}$$

$= 15 \text{ ms}^{-1}$

Which of the following statements is correct?
  a) The time taken to travel between lamp post 1 and 2 is the same as the time taken to travel between lamp post 2 and 3.
  b) The time taken to travel between lamp post 1 and 2 is greater than the time taken to travel between lamp post 2 and 3.
  c) The time taken to travel between lamp post 2 and 3 is greater than the time taken to travel between lamp post 1 and 2.
  d) The distance between the lamp posts is changing.

The car is accelerating which means that its speed is increasing. This means that as time = distance / speed and speed is increasing it will take less time to travel the distance between the lamp posts.

a) This is incorrect as the speed is changing therefore the time between lamp posts will change.

b) This is the correct answer.

c) This is incorrect as it would mean that speed is getting smaller.

d) This answer is here to throw you off track. The distance between the lamp posts is fixed.

1.3) A go cart is travelling down a hill with constant acceleration.
At time t = 3 seconds it has a velocity of 12 ms^{-1}. At time t = 6 seconds it has a velocity of 24 ms^{-1}.
  a) Calculate the acceleration of the go cart.
  b) Calculate the distance travelled during this time.

a) $\text{acceleration} = \frac{\text{(change in) speed}}{\text{(change in) time}}$

$= \frac{12 \text{ ms}^{-1}}{3 \text{ s}}$

$= \frac{(24 - 12) \text{ ms}^{-1}}{(6 - 3) \text{ s}}$

$= 4 \text{ ms}^{-2}$

b) For this step it is important to draw a speed time graph.

The area under the graph is the distance travelled. In this case it is the area of a triangle and the area of a square. We are only interested in the times listed between 3 and 6 seconds.

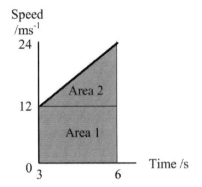

area 1 = 12 ms⁻¹ x (6 – 3) s
        = 36 m
area 2 = ½ x (24 – 12) ms⁻¹ x (6 – 3) s
        = 18 m

total area = area 1 + area 2
           = 54 m

The distance travelled during the 3 seconds the cart is accelerating is 54 m.

1.4) A motorbike is accelerating from rest. It reaches a velocity of 35 ms⁻¹ after 12 seconds and then continues at this speed for 20 seconds before slowing to a stop 40 seconds after it began to move.
     a) Calculate the initial acceleration.
     b) Calculate the final acceleration.
     c) Calculate the total distance travelled.
As this question is a little complex it is a good idea to draw it before we begin the question.

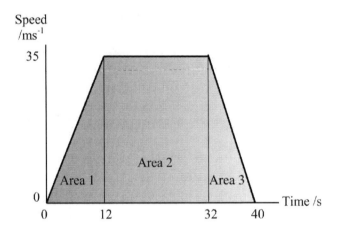

a) acceleration = (change in) speed / (change in) time

                = final speed – initial speed / final time – initial time

                = (35 – 0) ms⁻¹ / (12 – 0) s

                = 2.92 ms⁻²

b) acceleration = $\frac{\text{(change in) speed}}{\text{(change in) time}}$

$= \frac{\text{final speed} - \text{initial speed}}{\text{final time} - \text{initial time}}$

$= \frac{(0-35) \text{ ms}^{-1}}{(40-32) \text{ s}}$

$= -4.38 \text{ ms}^{-2}$

c) The total distance travelled is just the area under the speed-time graph.

area under the graph = area 1 + area 2 + area 3

area 1 = ½ x 12 s x 35 ms⁻¹
      = 210 m

area 2 = 20 s x 35 ms⁻¹
      = 700 m

area 3 = ½ x 8 s x 35 ms⁻¹
      = 140 m

area under the graph = 1,050 m
total distance travelled = 1,050 m

1.5) A sports fan is viewing a race through some binoculars. She sees the flash from the starting gun and then, after 4 seconds, she hears the sound. The speed of sound is 340 ms⁻¹.
 a) Complete the table.
 b) How far away is the sports fan from the racers?
 c) What assumptions have been made in this calculation?

a) Row 2: time = $\frac{\text{distance}}{\text{speed}}$

$= \frac{680 \text{ m}}{340 \text{ ms}^{-1}}$

= 2 s

Row 3: distance = time x speed
              = 3 s x 340 ms⁻¹
              = 1,020 m

Row 4: distance = time x speed
              = 4 s x 340 ms⁻¹
              = 1,360 m

Speed of sound / ms⁻¹	Distance to travel / m	Time for sound to travel the distance / s
340	340	1
340	680	2
340	1,020	3
340	1,360	4

b) How far away is the sports fan from the racers?

distance = time x speed
= 4 seconds x 340 ms⁻¹
= 1,360 metres

c) What assumptions have been made in this calculation?

We have assumed that the sports fan has seen the flash from the starting pistol at the moment it was fired (we have also assumed that the speed of light is infinite. It isn't, but it is very, very, big).

1.6) An empty bus accelerates from a bus stop and travels along a street to the next bus stop. There it picks up 7 people and accelerates from the bus stop. The engine is producing the same amount of force to move the bus however the bus accelerates more slowly. Why?

Something happens:
More people have entered the bus.

Which has an effect:
The mass of the bus and people has increased.

Which means:
The same force is trying to accelerate this larger mass (acceleration = force / mass).

Therefore:
The acceleration of the bus is smaller and the bus accelerates more slowly.

Final answer:
More people have entered the bus. The combined mass of the bus and people has increased. The same force is trying to accelerate this larger mass (acceleration = force / mass). The acceleration of the bus is smaller and the bus accelerates more slowly.

1.7) The acceleration due to gravity on the moon is 1.62 ms⁻². A golfer from Earth has travelled to the moon to take part in a golf tournament. On Earth he is normally able to hit a golf ball 240 yards. He hits the ball on the moon and is surprised at how far the ball travels. Explain this.

Something happens:
The golfer hits the ball.

Which has an effect:
The ball then travels upwards and away from the golfer at the speed that the golfer would expect on Earth. The acceleration due to gravity is much smaller on the moon.

Which means:
This means that the rate at which the golf ball travels downward towards the moon is much smaller than it would be on planet Earth but it still travels away at the same speed.

Therefore:
The ball will travel much farther before hitting the ground.

Final Answer:
The golfer hits the ball. The ball then travels upwards and away from the golfer at the speed that the golfer would expect on Earth. The acceleration due to gravity is much smaller on the moon. This means that the rate at which the golf ball travels downward towards the moon is much smaller than it would be on planet Earth but it still travels away at the same speed. Therefore, the ball will travel much farther before hitting the ground.

1.8) A car is broken down. A pedestrian offers to help. The driver believes that if the car can be pushed to a speed of 2 ms^{-1} he will be able to start it. The pedestrian agrees to push the car but asks the driver to remove all of his heavy shopping first. Why is this a good idea?

Something happens:
The shopping is removed from the car.

Which has an effect:
The total mass of the car will decrease.

Which means:
As acceleration = force / mass, the acceleration will be larger.

Therefore:
The car will reach the required speed more quickly.

Final Answer:
If the shopping is removed from the car. The total mass of the car will decrease. As acceleration = force / mass, the acceleration will be larger. The car will reach the required speed more quickly.

1.9) A girl has a mass of 40 kg. Calculate her weight on the following planets.
    Pluto, gravity = 0.9 ms^{-2}
    Venus, gravity = 9 ms^{-2}
    Earth, gravity = 10 ms^{-2}
    Jupiter, gravity = 25 ms^{-2}

weight = mass x gravity

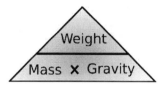

weight on Pluto = 40 kg x 0.9 ms^{-2}
              = 36 N

weight on Venus = 40 kg x 9 ms^{-2}
              = 360 N

weight on Earth = 40 kg x 10 ms^{-2}
              = 400 N

weight on Jupiter = 40 kg x 25 ms^{-2}
                = 1,000 N

1.10) The engine on a moving car develops a force of 2,000 N accelerating it forward. When the car has a speed of 30 ms^{-1} it no longer accelerates.
    a) Calculate the resistance force acting on the car
    b) What happens in the following circumstances?
        i) The resistive force acting on the car decreases.
        ii) The resistive force acting on the car increases.
        iii) The resistive force acting on the car remains the same.

a) force = acceleration x mass
resultant force = force forward + force backwards
            = 0 N

As there is no acceleration (it is moving at a constant speed) the value of the resultant force must equal zero.

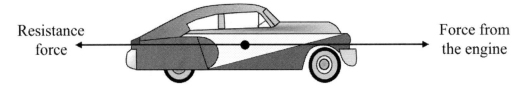

Force acting on car pushing it forward = 2,000 N.
Force acting on the car due to resistance = -2,000 N.

The negative value on the force tells us the direction of the force (it is in the opposite direction to the driving force from the engine).

b) What happens in the following circumstances?

i) The resistive force acting on the car decreases.

$$\text{acceleration} = \frac{\text{force}}{\text{mass}}$$

The resultant force will be positive (in the same direction as the car is moving) and therefore the acceleration will be positive and the car will accelerate and speed up.

ii) The resistive force acting on the car increases.

$$\text{acceleration} = \frac{\text{force}}{\text{mass}}$$

The resultant force will be negative (in the opposite direction to the movement of the car) and therefore the acceleration will be negative and the car will slow down.

iii) The resistive force acting on the car remains the same.

$$\text{acceleration} = \frac{\text{force}}{\text{mass}}$$

The resultant force will remain zero. There will be no acceleration or deceleration. The car will continue with constant speed.

1.11) Which of the following is true for motion in a circle?
   a) The speed is changing and the velocity is constant.
   b) The speed is constant and the velocity is changing.
   c) The acceleration is constant and the force is changing.
   d) The acceleration is changing and the force is constant.

For objects moving in a circular motion, only speed is a scalar quantity. This means that it has only a magnitude and no direction. Everything else is a vector quantity and changes direction as the object moves because the direction is always towards the centre.

The answer is B.

1.12) The following diagram shows a toy that is moving clockwise in a circle. Which of the following options is accurate?

   a) 1 is velocity, 2 is speed
   b) 2 is velocity, 3 is acceleration
   c) 3 is force, 4 is velocity
   d) 4 is velocity, 1 is speed

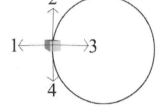

- As the toy is moving in a circular motion, the acceleration and force always point towards the centre. This would be arrow 3.

- The velocity is in the direction that the object is moving. As it is moving clockwise, arrow number 2 shows the velocity.

- Speed has no direction.

The answer is B.

Use the graph below to answer the next 4 questions.

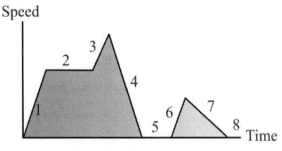

The above graph shows the motion of a cat playing with a toy. Distance is measured from a point in the centre of the room.

1.13) Which list identifies all of the areas where the cat was definitely accelerating?
   a) 1,3,4,6,7
   b) 2,5,8
   c) 1,3,6
   d) 4,7
   e) None of the above

The cat is undergoing acceleration at any time when the speed is changing. Remember that deceleration (where the object slows down) is still a type of acceleration. There is always an acceleration when the speed or velocity is changing.

- The answer is A.

1.14) Which list identifies the times when the cat was at its original location?
   a) 2,5,8
   b) 1,3,6
   c) 5,8
   d) 4,7
   e) None of the above

As it is a speed time graph there is no way to tell the position of the cat, only the rate at which it is travelling.

- The answer is E.

1.15) At one point the cat is running in a circle at constant rate about the measurement point. When was this?
   a) 1
   b) 2
   c) 5
   d) 8
   e) None of the above

A cat moving in a circle will have a constant speed that is not zero. Section 2 is the only time when this happens.

- The answer is B.

1.16) When was the deceleration of the cat the greatest?
   a) 1
   b) 3
   c) 4
   d) 7
   e) None of the above

Acceleration can be called deceleration when the object is slowing down. We are looking for the line with the steepest gradient when the cat is going from a fast to a slow speed. This is section 4.

- The answer is C.

Bonus Questions 1.1: Speed, Distance, Force and Weight

Use gravity = 10 ms^{-2} if it is not defined in the question.

1.1.1) A bicycle accelerates from rest at a rate of 3 ms^{-2} for 5 seconds. It then travels at a constant speed for a period of 20 seconds. After this period, it decelerates at a rate of 2 ms^{-2} for 6 seconds. Calculate the following.
    a) The speed that the bicycle has after its initial acceleration.
    b) The distance the bicycle travels at a constant speed.
    c) The final speed of the bicycle.

1.1.2) A runner accelerates at a constant rate from rest to a speed of 10 ms^{-1} in a time of 2.8 seconds.
    a) What is the acceleration of the runner?
    b) How far does the runner travel? Draw a graph to calculate the answer.

1.1.3) A screwdriver is dropped by a workman who is working at the top of a tall building. The screwdriver falls for 4 seconds before hitting the ground.
    a) What speed does the screwdriver have as it hits the ground?
    b) What assumptions have you made?

1.1.4) A diver jumps from the diving board to dive into the water. On their jump they accelerate from a point where they are stationary to a speed of 3 ms^{-1} in a time of 0.4 s. You may assume that air resistance is negligible.
    a) What is their acceleration during their take off jump?
    b) What is their acceleration the moment that their feet are no longer touching the jumping board?

1.1.5) A car accelerates from the lights with an acceleration of 1.6 ms^{-2}. It accelerates at this rate for a total of 12 seconds. After these 12 seconds have passed the car begins to slow down at a constant rate of 1.2 ms^{-2}. This car continues to slow until it is stationary at the next set of lights.
    a) What was the maximum speed of the car?
    b) How long did it take the car to stop from its maximum speed?
    c) How far apart are the lights?

1.1.6) A child is on a skateboard at the top of a ramp. They accelerate down the ramp at a constant rate of 2.3 ms^{-2}. They reach the bottom of the ramp in a time of 2.6 seconds. Once they reach the bottom of the ramp their speed decreases at a rate of 0.7 ms^{-2} until they are stopped.
    a) What speed did they reach at the base of the ramp?
    b) How long was the ramp? (Hint: draw a graph.)
    c) How far did they travel from the base of the ramp before they stopped? (Hint: draw a graph.)

1.1.7) A golf ball is hit by a professional golfer. During the contact with the club the golf ball accelerates from 0 ms^{-1} to 56 ms^{-1} in a time of 0.002 seconds.
    a) What is the acceleration of the golf ball caused by contact with the golf club?
    b) Assuming that air resistance is negligible, what is the acceleration of the golf ball once it is no longer in contact with the golf club?

1.1.8) A truck is stopped by the side of the road. It begins to accelerate at a constant rate and reaches a speed of 12.5 ms^{-1} in a time of 6 seconds. The truck driver sees an accident in the road in front of him and immediately applies the brakes. The truck decelerates at a constant rate and stops from its maximum speed in a time of 2.5 seconds.
    a) What was the initial acceleration of the truck?
    b) What was the deceleration of the truck during the final stage?
    c) What was the acceleration of the truck during its final stage?

1.1.9) A tennis ball is being hit between two players on opposite sides of the court. Assuming air resistance is negligible, what is the acceleration of the tennis ball when it is not in contact with the tennis racket, ground or net?

1.1.10) A truck needs to accelerate to a speed of 12 ms^{-1} in a time of 5 seconds. If the truck has a mass of 4,000 kg what force must be provided? You may assume that frictional forces are negligible.

1.1.11) A small engine provides a force of 300 N. Calculate the acceleration that the following objects would experience if the engine was attached to them.
   a) A bus, mass = 5,000 kg.
   b) A truck, mass = 3,000 kg.
   c) A motorcycle and driver, mass = 200 kg.

1.1.12) A force is acting on a car of mass 2,000 kg. The car accelerates from 3 ms^{-1} to 9 ms^{-1} in a time of 2.3 seconds. What is the force acting on the car?

1.1.13) A bicycle accelerates from rest at a rate of 3 ms^{-2}. If the mass of the cyclist and the bike is equal to 75 kg calculate the value of the force that is needed to accelerate the bike at this rate. You may ignore frictional forces.

1.1.14) A set of brakes is able to provide a braking force of 500 N. They are applied to a selection of different vehicles. Calculate the rate at which the following vehicles would slow down when the brakes were applied.
   a) A truck, mass = 3,400 kg.
   b) A car, mass = 2,500 kg.
   c) A small car, mass = 1,000 kg.

1.1.15) A new type of lubricating grease is being used in an old pulley mechanism. The 30 kg mass that was being lifted initially accelerated before at 0.6 ms^{-2}. Now it accelerates at 0.8 ms^{-2}. What is the extra force that is being used to lift the mass?

1.1.16) A man is pushing his car with a force of 300 N. The car is accelerating at 0.1 ms^{-2}. He stops and removes some items from the trunk of the car. Now when he pushes the car it accelerates at a rate of 0.101 ms^{-2}. What was the mass of the items that he removed?

1.1.17) A toy car is being pulled on a string. It has a mass of 100 g. It is pulled with a force of 0.3 N.
   a) What is the rate at which it accelerates?
   b) Why does it accelerate more slowly when it is being pulled over sand?

1.1.18) A new design of fuel injector is being tested. It should have the effect of producing another 1,000 N of force from the engine of the vehicle. Calculate the change that this will have on the acceleration of the following vehicles.
   a) A small car, mass = 1,200 kg.
   b) A small truck, mass = 4,000 kg.
   c) A large truck, mass = 8,000 kg.

1.1.19) Calculate the weight of a 1 kg mass in the following places.
   a) Earth, gravity = 10 ms^{-2}
   b) The moon, gravity = 1.62 ms^{-2}
   c) Pluto, gravity = 0.42 ms^{-2}

1.1.20) I have a mass of 70 kg. What would my weight be in the following places?
   a) Earth, gravity = 10 ms^{-2}
   b) The surface of the sun, gravity = 274.13 ms^{-2}
   c) The surface of a neutron star, gravity = 1.8 x 10^8 ms^{-2}

1.1.21) Calculate the mass of a rock that weighs 120,000 N on the surface of Jupiter.
(Gravity on Jupiter = 25.95 ms^{-2})

1.1.22) A large hammer on earth has a mass of 3 kg. What mass does it have in the following places?
   a) The moon, gravity = 1.62 ms^{-2}
   b) Deep space, gravity = 0 ms^{-2}

1.1.23) How much would a 4 kg cat weigh in the following places?
   a) Earth, gravity = 10 ms^{-2}
   b) Jupiter, gravity = 25.95 ms^{-2}
   c) The surface of the sun, gravity = 274.13 ms^{-2}

1.1.24) The International Space Station has a mass of 375,000 kg. It is in orbit around the Earth at a height of about 400 km from the surface of the Earth. The weight of the International Space Station is 3,337,500 N.
   a) Calculate the value of the gravitational acceleration at this height.
   b) Express this as a multiple of g, where g is the value of gravity at the Earth's surface.
   c) Calculate the weight of a 5,000 kg object that is in deep space far away from any galaxies.

1.1.25) If I weighed myself on the moon my weight would be 138 N. What is my mass?
(Gravity on the moon =1.62ms^{-2})

1.1.26) What is the weight of a 20 kg object that has just been released and is in free fall? Ignore the effects of any air resistance.

1.1.27) A runner is 10 m from a lamp post and is running away from it at a rate of 7 ms^{-1}. Calculate the distance the runner will have from the lamppost after;
   a) 20 seconds
   b) 5 minutes

1.1.28) A sprinter competing at the Olympic level runs 100 m in a time of 9.80 seconds.
   a) What average speed were they travelling?
   b) A long distance runner runs 400 m in a time of 58 seconds. How much faster is the sprinter travelling compared to the long distance runner?

1.1.29) Four homing pigeons are released at the same time and all travel to the same final point. They travel with the following speeds;
   Spiky travels at 8 ms^{-1} Fluffy travels at 5 ms^{-1}
   Stripy travels at 8.4 ms^{-1} Bernard travels at 9 ms^{-1}
Assuming that they are all exactly 25 km from their destination, which pigeon will arrive first and how long will it take to the nearest second?

1.1.30) For an object moving in a circle, which of the following is accurate?
   a) Force is away from the centre. Acceleration is away from the centre.
   b) Force is towards the centre. Acceleration is away from the centre.
   c) Force is away from the centre. Acceleration is towards the centre.
   d) Force is towards the centre. Acceleration is towards the centre.

1.1.31) Write, in words, Newton's first law and Newton's second law.

Use the graph below to answer the next 4 questions.

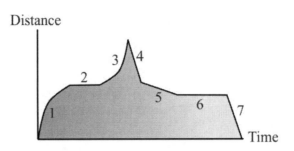

A student was testing their remote controlled car in the playground of their school. The graph above shows the record of the motion. The distances were measured from the student.

1.1.32) Which parts of the graph show a region where the radio controlled car is accelerating or decelerating?
   a) 1,3,4,5,7
   b) 4,5,6,7
   c) 1,2,3
   d) 1,3
   e) None of the above

1.1.33) Which parts of the graph show all of the regions where the car is moving with constant speed?
   a) 1,3
   b) 2,4,5,6,7
   c) 1,2,4,5,7
   d) 4,5,7
   e) None of the above

1.1.34) Choose a possible set of explanations that could both possibly describe part 2 of the graph above.
   a) The car could be moving in a circle about the student or standing still.
   b) The car could be moving away from the student with constant speed or standing still.
   c) The car could be moving towards the student with increasing acceleration or moving with constant speed.
   d) The car could be moving away from the student with constant acceleration or moving with constant speed.

1.1.35) Identify the areas of the graph where the car is changing speed or could be stationary.
   a) 1,3,4,7
   b) 1,2,3,6
   c) 2,4,5,6,7
   d) 2,6
   e) None of the above

Bonus Questions 1.2: Speed, Distance, Force and Weight

Use gravity = 10 ms^{-2} if it is not defined in the question.

1.2.1) A small car is accelerating at a rate of 2 ms^{-2} over a time of 6 seconds. It is accelerating from rest. After this initial acceleration, the car travels with a constant speed for a period of 4 seconds. It then slows down at a constant rate of 1.2 ms^{-2} for a period of 8 seconds.
   a) What is the speed of the car after its initial period of acceleration?
   b) Over what distance does the car travel at a constant speed?
   c) What is the final speed of the car?

1.2.2) A small dog is running at a speed of 3 ms^{-1}. It accelerates at a constant rate to a speed of 8 ms^{-1} in a time of 4 seconds.
   a) What is the acceleration of the dog?
   b) How far does the dog travel in this time? (Hint: sketch the graph and calculate the area.)

1.2.3) A red ball is thrown from the top of a building. It falls for 6 seconds before hitting the ground.
   a) Calculate the speed that the ball will hit the ground.
   b) What assumptions have you made?
   c) How will the real speed with which the ball would hit the ground compare with the value in (a)?

1.2.4) A ball is bounced on the ground. As it hits the ground, it accelerates to a speed of 6 ms^{-1} from rest in a time of 0.01 s.
   a) What is its acceleration during this time?
   b) What is the value of the acceleration of the ball once it has lost contact with the ground? (You may assume that air resistance is negligible.)

1.2.5) A glass ball is placed at the top of an inclined plane. It accelerates down the plane at a constant rate of 0.1 ms^{-2} reaching the bottom after 8 seconds. Its speed then decreases at a rate of 0.05 ms^{-2} until it eventually stops.
   a) What was the maximum speed of the ball?
   b) How long is the inclined plane? (Hint: sketch the graph to find the area.)
   c) What is the total distance travelled by the ball?

1.2.6) A cyclist has an acceleration of 1.6 ms^{-2}. It accelerates at this rate from rest for 9 seconds. After this time the cyclist stops pushing down hard on the pedals to rest and begins to slow down at a rate of 0.9 ms^{-2}. The cyclist continues to slow to an eventual stop.
   a) What was the maximum speed of the cyclist?
   b) How long does it take for the cyclist to slow from its maximum speed to a stop?
   c) What is the distance travelled by the cyclist from start to finish?

1.2.7) A pool ball is hit by a player during a game. While it is in contact with the cue, the pool ball accelerates from 0 ms^{-1} to 7 ms^{-1} in a time of 0.015 seconds.
   a) What is the acceleration of the pool ball?
   b) What distance does it travel during the acceleration?

1.2.8) A bus is taking some tourists on holiday. The driver has to stop to change a wheel. When he begins to accelerate after changing the wheel, he accelerates at a constant rate for 4 seconds and reaches a speed of 9.8 ms^{-1}. The driver then sees a red light in front of him and he applies the brakes. The bus slows to a stop in a time of 5.8 seconds.
   a) What was the initial acceleration of the truck?
   b) What was the deceleration of the bus during the final stage?
   c) How would the acceleration of the bus compare with the deceleration when the bus is slowing down?

1.2.9) A ping pong ball is being hit across the table by a player. Assuming air resistance is negligible, state the acceleration of the ball after the player hits it and it is no longer in contact with the paddle.

1.2.10) A small delivery van is pulling onto a busy highway. To do this safely it must reach a speed of 22 ms^{-1} in a time of 8 seconds. The van has a mass of 2,500 kg. What force must be provided by the engine? You may assume that frictional forces are negligible.

1.2.11) Calculate the force required to accelerate the following objects at a rate of 3 ms^{-2}.
   a) A coach, mass = 5,500 kg.
   b) A boat, mass = 25,000 kg.
   c) A dune buggy, mass = 450 kg.

1.2.12) A small car of mass 1,200 kg accelerates from 2 ms^{-1} to 10 ms^{-1} in a time of 4 seconds. What is the value of the force acting on the car?

1.2.13) A skateboarder accelerates from rest at a rate of 3 ms^{-2}. If the mass of the person and the skateboard is equal to 60 kg, how much force is needed? You may ignore frictional forces.

1.2.14) A new type of tire is being tested on some different vehicles. The tire is able to generate a braking force of 1,200 N. Calculate the rate at which the following vehicles would slow down using the new tires.
   a) A truck, mass = 4,100 kg.
   b) A car, mass = 1,500 kg.
   c) A motorbike, mass = 200 kg.

1.2.15) An elevator's drive motor has recently been serviced and its performance has improved. Before being serviced, the 500 kg elevator initially accelerated at 0.2 ms^{-2}. Now it has been serviced, the elevator has an initial acceleration of 0.25 ms^{-2}. What is the extra force that is being used to lift the elevator?

1.2.16) A delivery driver reaches his destination. He unloads his cargo and does not put anything else into the truck. He accelerates away from the depot. What happens to the acceleration for the same amount of force from the engine? Why?

1.2.17) A cart is pushed along a smooth surface. The cart has a mass of 30 kg and is being pushed with a force of 25 N.
   a) What is the rate at which it accelerates?
   b) What would happen to the acceleration if the surface was rough rather than smooth? Why?

1.2.18) A new design of engine has been built that will add another 3,000 N to the driving force on a vehicle. For the following vehicles, calculate the change that this will have on the acceleration.
   a) A car, mass = 1,450 kg.
   b) A small truck, mass = 3,500 kg.
   c) A bus, mass = 6,000 kg.

1.2.19) Calculate the weight of a 50 kg accountant in the following places.
   a) Earth, gravity = 10 ms^{-2}
   b) Jupiter, gravity = 25.95 ms^{-2}
   c) Mars, gravity = 3.7 ms^{-2}

1.2.20) A car has a weight of 17 kN. What would its weight be in the following places?
   a) Venus, gravity = 8.9 ms^{-2}
   b) Saturn, gravity = 10.44 ms^{-2}
   c) The surface of a neutron star, gravity = 1.8 x 10^8 ms^{-2}

1.2.21) The mars rover has a mass of 175 kg. Calculate the mass of the rover in the following places.
   a) Earth, gravity = 10 ms^{-2}
   b) Venus, gravity = 8.9 ms^{-2}
   c) Mars, gravity = 3.7 ms^{-2}

1.2.22) Calculate the mass of a rock that weighs 270 N on the moon.
(Gravity on the Moon = 1.62 ms^{-2})

1.2.23) How much would a dog that weighs 50 N on Earth weigh in the following places? Take Earth's gravity to be 10 ms^{-2}.
   a) Jupiter, gravity = 25.95 ms^{-2}
   b) Mars, gravity = 3.7 ms^{-2}

1.2.24) If I weighed myself on Jupiter, my weight would be 1,500 N. What is my mass?
(gravity on the Jupiter = 25.95 ms^{-2})

1.2.25) Calculate the weight of a diver with a mass of 68 kg who had just jumped from a diving board. The diver is falling and feels weightless. Ignore the effects of any air resistance.

1.2.26) A jogger is running towards a tree. They will run past the tree and continue in a straight line. They are initially 30 m from the tree. They are running at a rate of 7 ms^{-1}. Calculate the distance the jogger will have from the tree after;
   a) 30 seconds
   b) 2 minutes

1.2.27) A runner competing at the Olympic level runs 400 m in a time of 43.0 seconds.
   a) What was the average speed they were travelling at?
   b) A long distance runner runs 400 m in a time of 58 seconds. How much faster is the first runner travelling compared to the long distance runner?

1.2.28) 4 fish are released at the same time and all travel in a straight line to the same final point. They travel with the following speeds:

   Goldy travels at 0.5 ms^{-1} Scales travels at 0.45 ms^{-1}
   Herbert travels at 0.64 ms^{-1} Elf travels at 0.59 ms^{-1}

Assuming that they are all exactly 400 m from their destination, which fish will arrive first and how long will it take from the time the first fish arrives to the arrival of the second fish?

1.2.29) For an object moving in a circle, which of the following options is accurate?
   a) Velocity is changing and acceleration is changing.
   b) Speed is constant and acceleration is constant.
   c) Velocity is constant and acceleration is changing.
   d) Speed is changing and acceleration is constant.

1.2.30) Write, in words, Newton's first law and Newton's third law.

Use the graph below to answer the next 4 questions.

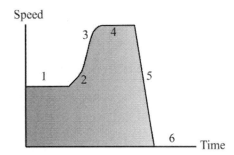

The graph above describes the journey of a car.

1.2.31) Identify the times when the car was decelerating.
   a) 1,4,6
   b) 2,3,5
   c) 3,5
   d) 5
   e) None of the above

1.2.32) Identify when the car was travelling with a constant speed.
   a) 1,4,6
   b) 5
   c) 1,5
   d) 6
   e) None of the above

1.2.33) When is the car travelling the fastest?
   a) 1,4,6
   b) 2,3,5
   c) 3,5
   d) 4
   e) None of the above

1.2.34) Which list shows all of the times that the car was accelerating?
   a) 1,4,6
   b) 2,3,5
   c) 2,3
   d) 2
   e) None of the above

2. Equations of Motion and More Complex Movement

Ugly and Amazing Teapots

I was cleaning out my old Aunt Bessy's attic and I found literally hundreds of hideous teapots as far as the eye could see. There were big teapots, small teapots, teapots shaped like chickens and other teapots so intricate you would never believe me if I told you. So I took all my money and opened up a tourist attraction from my shed. This is what I put on the brochure.

<u>V</u>iew <u>ugly</u> <u>and</u> <u>a</u>mazing <u>t</u>eapots!

<u>S</u>ee <u>u</u>gly <u>t</u>eapots <u>and</u> <u>have</u> <u>a</u> <u>t</u>ea <u>too</u>!

<u>To</u> <u>a</u>ll schools <u>and</u> <u>you</u> <u>too</u>,
<u>v</u>isit <u>to</u>day

<u>to</u> <u>s</u>ee <u>us</u> <u>and</u> <u>v</u>iew <u>t</u>eapots.

<u>V</u>iew <u>u</u>gly <u>and</u> <u>a</u>mazing <u>t</u>eapots!

<u>S</u>ee <u>u</u>gly <u>t</u>eapots <u>and</u> <u>have</u> <u>a</u> <u>t</u>ea <u>too</u>!

<u>To</u> <u>a</u>ll <u>s</u>chools <u>and</u> <u>you</u> <u>too</u>,
<u>v</u>isit <u>to</u>day

<u>to</u> <u>s</u>ee <u>u</u>s <u>and</u> <u>v</u>iew <u>t</u>eapots.

<u>V</u>iew <u>u</u>gly <u>and</u> <u>a</u>mazing <u>t</u>eapots!	v u + at	**1**
<u>s</u>ee <u>u</u>gly <u>t</u>eapots <u>and</u> <u>have</u> <u>a</u> <u>t</u>ea <u>too</u>!	s ut + ½at²	**2**
<u>To</u> <u>a</u>ll <u>s</u>chools <u>and</u> <u>you</u> <u>too</u>, <u>v</u>isit <u>to</u>day	2as + u² v²	**3**
<u>to</u> <u>s</u>ee <u>u</u>s <u>and</u> <u>v</u>iew <u>t</u>eapots.	2s (u + v)t	**4**

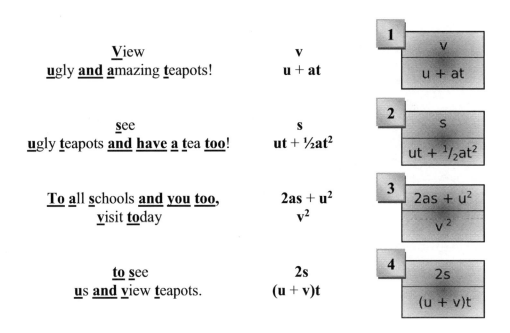

These are known as the SUVAT equations.

s = displacement
u = initial velocity
v = final velocity
a = acceleration
t = time

So how do you know which equation to use? The answer is always the same…

1. Write down what you know from the question. i.e. you have s,u,v and want to find a
2. Find the equation that has these terms i.e. $2as + u^2 = v^2$
3. Rearrange the equation to find what you want i.e. $a = (v^2 - u^2) / 2s$

The Square Method for Tricky Equations

You can use the square method to make solving equations more simple. This approach works well for people who are not confident rearranging equations but if you find this easy already, then use the method you are used to.

This method is used when we have two equations that must be equal to each other. The difference between the triangle method and the square method is one of complexity. Use the triangle method when you have very simple equations with no addition or subtraction. You can use the square method when the equation becomes more complex or has any addition or subtraction.

I will use the simplest square as an example, however these rules apply to even very complex equations.

- The first thing to know is that each side of the equation gets one line.

final velocity = initial velocity + (acceleration x time)

Now in order to solve for any term in this equation, there are only 2 simple rules.

Rule 1: Change the sign... Change the line
Cover over the quantity you are interested in. If there are any other quantities that are added to or subtracted from the quantity you are interested in they must be moved. So how are they moved?

- Reverse the sign (+ to - and vice versa) of anything added to or subtracted from the thing you are looking for and then place it on the other level of the square.

Rule 2: Put it on the bottom
If the quantity you are interested in is on the top of the square, then turn the square upside down for the calculation. If it is on the bottom of the square, then leave it as it is.

Let's use these rules to solve for 'a'.

Rule 1: When you cover up the 'a' you see that 'u' is added to it and it is multiplied by 't'. You only have to move the 'u'. Change the sign and move it to the top line.

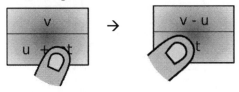

Rule 2: Make sure that the quantity you are looking for is on the bottom line. It already is so you do nothing here!

Now use the line to show division and you have your final equation!

$$\text{acceleration} = \frac{\text{final velocity} - \text{initial velocity}}{\text{time}}$$

That was a very simple equation. Now I will use an example from another chapter to show how this method can become handy with more complex equations.

The equation I will use is: total momentum before = total momentum after.

$$\text{Total momentum before} = \text{Total momentum after}$$

The equation for momentum is: momentum = velocity x mass
This is an equation that can be easily solved using the triangle method.

You can use the square when you make the momentum before and after a collision or event equal. Remember that each side gets its own line. (i = initial, f = final)

This makes the square: mass_i x velocity_i = mass_f x velocity_f

I have picked momentum for a very specific reason. Often in momentum there will be two separate masses at the start and they will collide and stick together to form one final mass.

On the top row of the square, $m_i v_i$ is made of two contributions. The initial momentum of one mass ($m_1 v_1$) is added to the initial momentum of the other mass ($m_2 v_2$), the two masses then combine. It does not matter which mass is m_1 or m_2.

mass_1 x velocity_1 + mass_2 x velocity_2 = mass_f x velocity_f

I want to find m_1.

Rule 1: $m_2 v_2$ must become negative and be moved to the base of the square.

Rule 2: The quantity is on the top line of the square so flip the square upside down.

Now we have our final equation:

$$m_1 = \frac{(m_f \times v_f) - (m_2 \times v_2)}{v_1}$$

Let's try again. I want to find v_2.

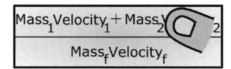

Rule 1: m_1v_1 must become negative and be moved to the base of the square.

$$\frac{Mass_2 \cdot v_2}{Mass_f Velocity_f - Mass_1 Velocity_1}$$

Rule 2: The quantity is on the top line of the square so flip the square upside down.

$$\frac{Mass_f Velocity_f - Mass_1 Velocity_1}{Mass_2 \cdot v_2}$$

$$v_2 = \frac{(m_f \times v_f) - (m_1 \times v_1)}{m_2}$$

One final example: I wish to find v_f.

Rule 1: Nothing else is added or subtracted on that line, so no change.

$$\frac{Mass_1 Velocity_1 + Mass_2 Velocity_2}{Mass_f \cdot v_f}$$

Rule 2: The quantity is on the bottom so there's no need to change anything.

$$v_f = \frac{(m_1 \times v_2) + (m_2 \times v_2)}{m_f}$$

Hopefully you will find this method easy to use so you never make mistakes rearranging more complex equations again!

Examples: Equations of Motion and More Complex Movement

Example 1) A toy car is controlled by a mobile phone. It accelerates from a speed of 0.5 ms^{-1} to 2 ms^{-1} in a distance of 3 metres. Calculate the acceleration of the car.

- You have initial speed (u), final speed (v) and displacement (s).
  You need acceleration (a) so use equation 3. This is the only equation with u,v,s and a.

$$\frac{2as + u^2}{v^2}$$

Rule 1: We want acceleration (a) so reverse the sign of the u^2 and put it on the other line.

$$\frac{2as}{v^2 - u^2}$$

Rule 2: Acceleration is located on the top part of the square so flip it upside down.

$$\frac{v^2 - u^2}{2as}$$

- Acceleration is now on the bottom and is only multiplied by some other values. Cover over the 'a' and this will reveal the equation to calculate. This gives us the final equation that we need.

$$a = \frac{v^2 - u^2}{2s}$$

$$\text{acceleration} = \frac{(\text{final velocity})^2 - (\text{initial velocity})^2}{2 \times \text{displacement}}$$

$$= \frac{(2 \text{ ms}^{-1})^2 - (0.5 \text{ ms}^{-1})^2}{2 \times 3 \text{ m}}$$

$$= 0.625 \text{ ms}^{-2}$$

Example 2) What distance is required for an object traveling with an initial velocity of 10 ms^{-1} to reach a final velocity of 30 ms^{-1} if it has an acceleration of 4 ms^{-2}?

- You have initial velocity (u), final velocity (v) and acceleration (a).
  You need displacement (s) so use equation 3.

$$\frac{2as + u^2}{v^2}$$

Rule 1: We want displacement (s) so reverse the sign of the u² and put it on the other line.

$$\frac{2as}{v^2 - u^2}$$

Rule 2: Displacement is located on the top part of the square so flip it upside down.

$$\frac{v^2 - u^2}{2as}$$

- Displacement is now on the bottom and is only multiplied by some other values. Cover over the 's' and this will reveal the equation to calculate. This gives us the final equation that we need.

$$\text{displacement} = \frac{(\text{final velocity})^2 - (\text{initial velocity})^2}{2 \times \text{acceleration}}$$

$$= \frac{(30 \text{ ms}^{-1})^2 - (10 \text{ ms}^{-1})^2}{2 \times 4 \text{ ms}^{-2}}$$

$$= 100 \text{ m}$$

Example 3) A moped accelerated at a rate of 2.5 ms⁻² over a displacement of 24 m. If its final velocity is 14 ms⁻¹, calculate the initial velocity of the moped.

- You have acceleration (a), displacement (s), and the final velocity (v).
You need the initial velocity (u) so use equation 3.

$$\frac{2as + u^2}{v^2}$$

Rule 1: We want initial velocity (u) so reverse the sign of the 2as and put it on the other line.

$$\frac{u^2}{v^2 - 2as}$$

Rule 2: The initial velocity is located on the top part of the square so flip it upside down.

$$\frac{v^2 - 2as}{u^2}$$

- Initial velocity is now on the bottom so covering it up will give the equation you want. As it is squared, you need to take the square root of the right hand side to find the initial velocity.

$$u = \sqrt{(v^2 - 2as)}$$

initial velocity = $\sqrt{((\text{final velocity})^2 - 2(\text{acceleration} \times \text{displacement}))}$
  = $\sqrt{((14 \text{ ms}^{-1})^2 - (2 \times 2.5 \text{ ms}^{-2} \times 24 \text{ m}))}$
  = $\sqrt{76} \text{ m}^2\text{s}^{-2}$
  = 8.72 ms^{-1}

Example 4) Find the final velocity of a bicycle with initial velocity 8 ms^{-1} undergoing an acceleration of 5 ms^{-2} over a distance of 3 metres.

- You have the initial velocity (u), acceleration (a) and displacement (s).
  You need the final velocity (v) so use equation 3.

$$\frac{2as + u^2}{v^2}$$

- We want final velocity (v). As it is already on the bottom, cover up the 'v' to get your final equation. Don't forget though that it is v^2 so you must take the square root to find the answer.

$$v = \sqrt{(2as + u^2)}$$

final velocity = $\sqrt{((2 \times \text{acceleration} \times \text{displacement}) + (\text{initial velocity})^2)}$
  = $\sqrt{((2 \times 5 \text{ ms}^{-2} \times 3 \text{ m}) + (8 \text{ ms}^{-1})^2)}$
  = $\sqrt{94} \text{ m}^2\text{s}^{-2}$
  = 9.70 ms^{-1}

Example 5) An object accelerates at a rate of 0.9 ms^{-2} for a period of 3 seconds. It has a final displacement of 19 metres. What is the initial velocity of the object?

- You have acceleration (a), time (t) and displacement (s).
  You need the initial velocity (u) so use equation 2.

$$\frac{s}{ut + \frac{1}{2}at^2}$$

Rule 1: We want the initial velocity (u) so reverse the sign of the ½at² and put it on the other line.

$$\frac{s - \tfrac{1}{2}at^2}{ut}$$

Rule 2: As 'u' is on the bottom, cover it over to find the equation you need to use.

$$u = \frac{s - \tfrac{1}{2}at^2}{t}$$

$$\text{initial velocity} = \frac{\text{displacement} - (1/2 \times \text{acceleration} \times \text{time}^2)}{\text{time}}$$

$$= \frac{19 \text{ m} - (1/2 \times 0.9 \text{ ms}^{-2} \times (3 \text{ s})^2)}{3 \text{ s}}$$

$$= 4.98 \text{ ms}^{-1}$$

Example 6) A squirrel starts with an initial velocity of 2 ms^{-1}. After a period of 5 seconds it has a displacement of 12 m. What is the value of the acceleration of the squirrel?

- You have initial velocity (u), time (t) and displacement (s).
 You need acceleration so use equation 2.

$$\frac{s}{ut + \tfrac{1}{2}at^2}$$

Rule 1: We want the acceleration (a) so reverse the sign of the ut and put it on the other line.

$$\frac{s - ut}{\tfrac{1}{2}at^2}$$

Rule 2: As 'a' is on the bottom, covering it over will now reveal the equation you need to use.

$$a = \frac{s - ut}{\tfrac{1}{2}t^2}$$

$$\text{acceleration} = \frac{\text{displacement} - (\text{initial velocity} \times \text{time})}{(1/2 \times \text{time}^2)}$$

$$= \frac{12 \text{ m} - (2 \text{ ms}^{-1} \times 5 \text{ s})}{\tfrac{1}{2} \times (5 \text{ s})^2}$$

$$= 0.16 \text{ ms}^{-2}$$

Example 7) A marble is dropped from rest from a height of 10 metres above a platform. It falls with an acceleration of -10 ms^{-2} for a period of 2 seconds.
a) What is its displacement after this time from its original position?

You have initial velocity (u), acceleration (a) and time (t). You need to find displacement (s) so use equation 2.

Because the initial velocity is zero, this removes the 'ut' to make the equation much simpler.

$$\boxed{\begin{array}{c} s \\ ut + \tfrac{1}{2}at^2 \end{array}} \rightarrow \boxed{\begin{array}{c} s \\ \tfrac{1}{2}at^2 \end{array}}$$

displacement = initial velocity x time + ½ x acceleration x time^2
= 0 ms^{-1} x 2 s + ½(-10 ms^{-2} x (2 s)^2)
= -20 m

b) Calculate the final height of the marble from the platform.

Height from platform = -20 m – (-10 m)
= -10 m
The marble has dropped 20 metres so it is now 10 metres below the platform.

c) Repeat the previous calculations with an initial velocity of 8 ms^{-1}.

displacement = initial velocity x time x ½ x acceleration x time^2
= 8 ms^{-1} x 2 s + ½(-10 ms^{-2} x (2 s)^2)
= -4 m

$$\boxed{\begin{array}{c} s \\ ut + \tfrac{1}{2}at^2 \end{array}}$$

Height from platform = -4 m – (-10 m)
= 6 m
The marble has dropped 4 metres so it is now 6 metres above the platform.

d) What is the significance of the negative sign in front of the 10 ms^{-2}?

The negative sign gives the direction, in this case downwards. (Positive values are upwards.) Make sure you pay attention to the sign!

Example 8) A small sponge ball is placed on a chute by a child. The chute has a length of 2.4 metres and the ball accelerates at a constant rate of 4 ms^{-2}. If the ball starts from rest, calculate the time required for the ball to travel the length of the chute.

You have displacement (s), acceleration (a) and initial velocity (u). You need to find time (t) so use equation 2.
As the initial velocity is zero, remove the 'ut' from the equation.

$$\boxed{\begin{array}{c} s \\ ut + \tfrac{1}{2}at^2 \end{array}}$$

time^2 = $\dfrac{\text{displacement}}{(1/2 \text{ x acceleration})}$

= $\dfrac{2.4 \text{ m}}{\text{½ x 4 ms}^{-2}}$

time = $\sqrt{1.2}$ s
= 1.10 s

Example 9) A cat is pacing along a road at a rate of 0.2 ms^{-1} it begins to accelerate at a rate of 1 ms^{-2} toward a leaf on the road. If it accelerated for a period of 8 seconds, what was the displacement of the cat from its original position?

You have initial velocity (u), acceleration (a) and time (t). You need displacement (s) so use equation 2.

displacement = (initial velocity x time) + ½(acceleration x time^2)
= (0.2 ms^{-1} x 8) + ½(1 ms^{-2} x 8^2)
= 33.6 m

$$\frac{s}{ut + \tfrac{1}{2}at^2}$$

Example 10) A moped travels through a displacement of 2 km in a time of 5 minutes.
a) If its final velocity is 10 ms^{-1} what is its initial velocity?

2 km = 2,000 m   5 minutes = 5 x 60 s min^{-1}
                              = 300 s

$$\frac{2s}{(u+v)t}$$

To separate out the displacement (u) from this equation, you need to multiply out the time first.

$$\frac{2s}{ut + vt}$$

Rule 1: Now you can put the 'vt' on the other line and cover over 't'.

$$\frac{2s - vt}{ut} \quad \rightarrow \quad u = \frac{2s - vt}{t}$$

initial velocity = $\dfrac{2 \times \text{displacement} - (\text{final velocity} \times \text{time})}{\text{time}}$

= $\dfrac{2 \times 2{,}000 \text{ m} - (10 \text{ ms}^{-1} \times 300 \text{ s})}{300 \text{ s}}$

= 3.33 ms^{-1}

b) What are you assuming?

In this question you are assuming that the acceleration is constant.

Example 11) A train has an initial velocity of 15 ms⁻¹ and a final velocity of -5 ms⁻¹.
a) If it has a final displacement of 300 m calculate the length of time for which it has travelled.

You have displacement (s), time (t) and final velocity (v). You need initial velocity (u) so use equation 4.

As time (t) is on the bottom and multiplies everything, you can just cover it over to find the equation you need.

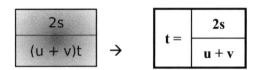

$$\text{time} = \frac{2 \times \text{displacement}}{\text{initial velocity} + \text{final velocity}}$$

$$= \frac{2 \times 300 \text{ m}}{15 \text{ ms}^{-1} + (-5 \text{ ms}^{-1})}$$

$$= 60 \text{ s}$$

b) What are you assuming?
You are assuming that the acceleration is constant.

Example 12) A truck is travelling down a lane. It has an initial velocity of 2 ms⁻¹ and a final velocity of 2.2 ms⁻¹. It travels through a displacement of 1.05 km. Calculate the time for which the truck was travelling.

You have initial velocity (u), final velocity (v) and displacement (s). You need to find time (t) so use equation 4.

displacement in metres = 1.05 km x 1,000 m km⁻¹
= 1,050 m

As time (t) is on the bottom and multiplies everything, you can just cover it over to find the equation you need.

$$\text{time} = \frac{2 \times \text{displacement}}{\text{initial velocity} + \text{final velocity}}$$

$$= \frac{2 \times 1,050 \text{ m}}{2 \text{ ms}^{-1} + 2.2 \text{ ms}^{-1}}$$

$$= 500 \text{ s}$$

Example 13) A hamster is travelling with a velocity of 3 ms⁻¹. It accelerates at a constant rate of 0.2 ms⁻² for a period of 2 seconds. Calculate its final velocity.

You have initial velocity (u), acceleration (a) and time (t).
You need final velocity (v) so use equation 1.

final velocity = initial velocity + (acceleration x time)
= 3 ms⁻¹ + (0.2 ms⁻² x 2 s)
= 3.4 ms⁻¹

Example 14) A mouse decelerates at a rate of 0.3 ms⁻² for 3 seconds. If it has a final velocity of 1.2 ms⁻¹, calculate its initial velocity.

You have acceleration (a), time (t) and final velocity (v). You need initial velocity so use equation 1.

Remember that deceleration is negative acceleration.

Rule 1: All you need to do is change the sign of 'at' and put it on the other line.

initial velocity = final velocity – (acceleration x time)
= 1.2 ms⁻¹ – (-0.3 ms⁻² x 3 s)
= 2.1 ms⁻¹

v
u + at

v - at
u

Example 15) How long does a bicyclist accelerate for if she accelerates at a rate of 0.6 ms⁻² from a velocity of 6 ms⁻¹ to 9.6 ms⁻¹?

You have acceleration (a), initial velocity (u) and final velocity (v). You need time (t) so use equation 1.

v
u + at

Rule 1: As you are looking for time (t) you must put the initial velocity (u) on the other line by changing the sign. Then you just cover up the 't' to find your equation.

time = $\dfrac{\text{final velocity - initial velocity}}{\text{acceleration}}$

= $\dfrac{9.6 \text{ ms}^{-1} - 6 \text{ ms}^{-1}}{0.6 \text{ ms}^{-2}}$

= 6 s

v - u
at

Example 16) A snowmobile accelerates down a ski slope at a rate of 5 ms^{-2} if its initial velocity was 3.4 ms^{-1} and its final velocity is 18.6 ms^{-1}, calculate the time for which the snowmobile was accelerating.

You have acceleration (a), initial velocity (u) and final velocity (v).
You need time (t) so use equation 1.

Rule 1: You are looking for time (t) so move the 'u' to the other line by changing the sign and then cover up the 't'.

time $= \dfrac{\text{final velocity} - \text{initial velocity}}{\text{acceleration}}$

$= \dfrac{18.6 \text{ ms}^{-1} - 3.4 \text{ ms}^{-1}}{5 \text{ ms}^{-2}}$

$= 3.04$ s

$t = \dfrac{v - u}{a}$

Questions: Equations of Motion and More Complex Movement

2.1) A boy is out exercising on a running track. He accelerates from a jog at 2 ms^{-1} to a run of 8 ms^{-1} over a total distance of 10 m. What is the acceleration of the boy during this time?

2.2) A cyclist has an initial velocity of 5 ms^{-1}. If the cyclist has a final speed of 12 ms^{-1} and she has accelerated at a rate of 3 ms^{-2} calculate the distance over which this has occurred.

2.3) A car is accelerating to a final velocity of 15 ms^{-1} over a displacement of 30 m. If it has an acceleration of 3 ms^{-2}, what was the initial velocity of the car?

2.4) A motorbike has an acceleration of 3 ms^{-2}. If their initial velocity was 8 ms^{-1} and they accelerate over a displacement of 38 m, what is their final velocity?

2.5) A toy car rolls down a ramp. It accelerates at a rate of 0.1 ms^{-2} for a period of 4 seconds until it reaches a displacement of 1.4 m from the top of the ramp. Calculate the initial velocity of the toy car.

2.6) Find the acceleration of a cheetah when it is pursuing its prey. It changes its velocity from 2 ms^{-1} over a distance of 7 m and a time of 1.5 s.

2.7) A fish travelling in a stream is initially at rest. Its final displacement was 45 cm. The fish experiences an acceleration of 5 ms^{-2}. Calculate the time required for the fish to travel through this distance.

2.8) A car is travelling along a small road. It has an initial velocity of 2 ms^{-1}. It experiences a constant acceleration of 1 ms^{-2} over a time period of 6 seconds. If it was initially next to a tree, what is its final displacement from the tree?

2.9) A bus accelerates from 10 ms^{-1} to 14 ms^{-1} in a time of 20 seconds. Calculate the displacement of the bus from its starting location.

2.10) A train accelerates to a velocity of 18 ms^{-1} in a time of 20 seconds. In this time period it covers a distance of 300 metres. Calculate the initial velocity of the train.

2.11) A dog has a displacement of 20 metres from its starting position. If it had an initial velocity of 1 ms^{-1} and a final velocity of 4 ms^{-1}, calculate the time for which it was travelling.

2.12) An eagle travels in a straight line through a distance of 4 km in a time of 10 minutes. If its initial velocity was 10 ms^{-1} and assuming constant acceleration, what is the value of its final velocity?

2.13) A small electric toy has an initial velocity of 0.1 ms^{-1}. It is accelerating down a ramp towards a jump. The toy accelerates at a rate of 0.05 ms^{-2} for a period of 8 seconds when it reaches the jump. Calculate the final velocity of the car as it hits the jump.

2.14) A moped is travelling along a smooth section of road. The moped accelerates at a rate of 1.2 ms^{-2} for a time of 6 seconds. It reaches a final velocity of 12.5 ms^{-1}. What was the initial velocity of the moped?

2.15) A cart is attached to the back of a small 4-wheel drive ATV. It accelerates at a rate of 1.3 ms^{-2}, changing its velocity from 1.6 ms^{-1} to 4.8 ms^{-1}. Calculate the time for which the cart is accelerated.

2.16) Calculate the acceleration required for a hawk to reach a final velocity of 20 ms^{-1} from an initial velocity of 4 ms^{-1} in a time of 3 seconds.

Answers: Equations of Motion and More Complex Movement

2.1) A boy is out exercising on a running track. He accelerates from a jog at 2 ms^{-1} to a run of 8 ms^{-1} over a total distance of 10 m. What is the acceleration of the boy during this time?

- You have initial velocity (u), final velocity (v) and displacement (s). You need acceleration (a) so use equation 3. This is the only equation with u,v,s and a.

$$\frac{2as + u^2}{v^2}$$

Rule 1: We want acceleration (a) so reverse the sign of the u^2 and put it on the other line.
Rule 2: Acceleration is on the top part so flip it upside down.
- Now cover over the 'a' to get the equation you need.

$$a = \frac{v^2 - u^2}{2s}$$

$$\text{acceleration} = \frac{(\text{final velocity})^2 - (\text{initial velocity})^2}{2 \times \text{displacement}}$$

$$= \frac{(8 \text{ ms}^{-1})^2 - (2 \text{ ms}^{-1})^2}{2 \times 10 \text{ m}}$$

$$= 3 \text{ ms}^{-2}$$

2.2) A cyclist has an initial velocity of 5 ms^{-1}. If the cyclist has a final speed of 12 ms^{-1} and she has accelerated at a rate of 3 ms^{-2} calculate the distance over which this has occurred.

- You have initial velocity (u), final velocity (v) and acceleration. You need displacement (s) so use equation 3.

Rule 1: We want displacement (s) so reverse the sign of the u^2 and put it on the other line.
Rule 2: Displacement is on the top part so flip it upside down.
- Now cover over the 's' to get the equation you need.

$$s = \frac{v^2 - u^2}{2a}$$

$$\text{displacement} = \frac{(\text{final velocity})^2 - (\text{initial velocity})^2}{2 \times \text{acceleration}}$$

$$= \frac{(12 \text{ ms}^{-1})^2 - (5 \text{ ms}^{-1})^2}{2 \times 3 \text{ m}}$$

$$= 19.8 \text{ m}$$

2.3) A car is accelerating to a final velocity of 15 ms⁻¹ over a displacement of 30 m. If it has an acceleration of 3 ms⁻², what was the initial velocity of the car?

- You have final velocity (v), displacement (s) and acceleration (a). You need the initial velocity (u) so use equation 3.

2as + u²
v²

  Rule 1: We want the initial velocity (u) so reverse the sign of the 2as and put it on the other line.
  Rule 2: The initial velocity is on the top part so flip it upside down.
- Now cover over the 'u' to get the equation you need. Remember that it is u² so you must take the square root to get the answer.

| u = | √(v² − 2as) |

initial velocity = √((final velocity)² − (2 × acceleration × displacement))
$$= \sqrt{(15 \text{ ms}^{-1})^2 - (2 \times 3 \text{ ms}^{-2} \times 30 \text{ m})}$$
$$= \sqrt{45} \text{ m}^2\text{s}^{-2}$$
$$= 6.7 \text{ ms}^{-1}$$

2.4) A motorbike has an acceleration of 3 ms⁻². If their initial velocity was 8 ms⁻¹ and they accelerate over a displacement of 38 m, what is their final velocity?

- You have acceleration (a), initial velocity (u) and displacement (s). You need final velocity so use equation 3.

2as + u²
v²

- We want the final velocity (v). To get the final equation, cover up the 'v' and don't forget that the final velocity is squared.

| v = | √(2as + u²) |

final velocity = √((2 × acceleration × displacement) + (initial velocity)²)
$$= \sqrt{(2 \times 3 \text{ ms}^{-2} \times 38 \text{ m}) + (8 \text{ ms}^{-1})^2}$$
$$= \sqrt{292} \text{ m}^2\text{s}^{-2}$$
$$= 17.1 \text{ ms}^{-1}$$

2.5) A toy car rolls down a ramp. It accelerates at a rate of 0.1 ms⁻² for a period of 4 seconds until it reaches a displacement of 1.4 m from the top of the ramp. Calculate the initial velocity of the toy car.

- You have acceleration (a), time (t) and displacement (s). You need the initial velocity (u) so use equation 2.

s
ut + ½at²

Rule 1: We want the initial velocity (u) so reverse the sign of the ½at² and put it on the other line.

$$\boxed{\begin{array}{c} s - \tfrac{1}{2}at^2 \\ \hline ut \end{array}}$$

Rule 2: As 'u' is on the bottom, cover it over to find the equation you need to use.

$$u = \boxed{\dfrac{s - \tfrac{1}{2}at^2}{t}}$$

$$\text{initial velocity} = \frac{\text{displacement} - (1/2 \times \text{acceleration} \times \text{time}^2)}{\text{time}}$$

$$= \frac{1.4\ \text{m} - (1/2 \times 0.1\ \text{ms}^{-2} \times (4\ \text{s})^2)}{4\ \text{s}}$$

$$= 0.15\ \text{ms}^{-1}$$

2.6) Find the acceleration of a cheetah when it is pursuing its prey. It changes its velocity from 2 ms⁻¹ over a distance of 7 m and a time of 1.5 s.

- You have initial velocity (u), displacement (s) and time (t). You need acceleration (a) so use equation 2.

$$\boxed{\begin{array}{c} s \\ \hline ut + \tfrac{1}{2}at^2 \end{array}}$$

Rule 1: We want the acceleration (a) so reverse the sign of the ut and put it on the other line. This reveals the equation you need to use.

$$\text{acceleration} = \frac{\text{displacement} - (\text{initial velocity} \times \text{time})}{(1/2 \times \text{time}^2)}$$

$$a = \boxed{\dfrac{s - ut}{\tfrac{1}{2}t^2}}$$

$$= \frac{7\ \text{m} - (2\ \text{ms}^{-1} \times 1.5\ \text{s})}{\tfrac{1}{2} \times (1.5\ \text{s})^2}$$

$$= 3.6\ \text{ms}^{-2}$$

2.7) A fish travelling in a stream is initially at rest. Its final displacement was 45 cm. The fish experiences an acceleration of 5 ms⁻². Calculate the time required for the fish to travel through this distance.

- You have initial velocity (u), displacement (s) and acceleration (a). You need time (t) so use equation 2.
- We know that the initial velocity is zero, so this allows us to make the equation simpler.

$$\boxed{\begin{array}{c} s \\ \hline ut + \tfrac{1}{2}at^2 \end{array}} \rightarrow \boxed{\begin{array}{c} s \\ \hline \tfrac{1}{2}at^2 \end{array}}$$

- We want time (t) so covering up the t and taking the root of the remaining values will give you your final equation to use.

$$t^2 = \frac{s}{\frac{1}{2}a}$$

- To make it even easier, remember that this is the same as:
- Remember that 45 cm = 0.45 m

$$t^2 = \frac{2s}{a}$$

$$\text{time}^2 = \frac{(2 \times \text{displacement})}{\text{acceleration}}$$

$$= \frac{(2 \times 0.45 \text{ m})}{5 \text{ ms}^{-2}}$$

$$\text{time} = \sqrt{0.18 \text{ s}^2}$$
$$= 0.42 \text{ s}$$

2.8) A car is travelling along a small road. It has an initial velocity of 2 ms⁻¹. It experiences a constant acceleration of 1 ms⁻² over a time period of 6 seconds. If it was initially next to a tree, what is its final displacement from the tree?

You have initial velocity (u), acceleration (a) and time (t).
You need displacement (s) so use equation 2.

$$\boxed{\frac{s}{ut + \frac{1}{2}at^2}}$$

displacement = initial velocity × time + ½(acceleration × time²)
$$= 2 \text{ ms}^{-1} \times 6 \text{ s} + \frac{1}{2}(1 \text{ ms}^{-2} \times (6 \text{ s})^2)$$
$$= 30 \text{ m}$$

2.9) A bus accelerates from 10 ms⁻¹ to 14 ms⁻¹ in a time of 20 seconds. Calculate the displacement of the bus from its starting location.

You have initial velocity (u), final velocity (v) and time (t).
You need displacement (s) so use equation 4.

$$\boxed{\frac{2s}{(u+v)t}}$$

Rule 2: You are looking for displacement (s) so you need to flip the square upside down.

$$\text{displacement} = \frac{(\text{initial velocity} + \text{final velocity}) \times \text{time}}{2}$$

$$s = \frac{(u+v)t}{2}$$

$$= \frac{(10 \text{ ms}^{-1} + 14 \text{ ms}^{-1}) \times 20 \text{ s}}{2}$$

$$= 240 \text{ m}$$

2.10) A train accelerates to a velocity of 18 ms^{-1} in a time of 20 seconds. In this time period it covers a distance of 300 metres. Calculate the initial velocity of the train.

You have final velocity (v), time (t) and displacement (s). You need initial velocity (u) so use equation 4.

To separate out the initial velocity (u) from this equation, you need to multiply out the time first.

$$\frac{2s}{(u+v)t} \rightarrow \frac{2s}{ut+vt}$$

Rule 1: Now you can put the 'vt' on the other line and cover over 'u'.

$$\frac{2s-vt}{ut} \rightarrow u = \frac{2s-vt}{t}$$

initial velocity $= \dfrac{2 \times \text{displacement} - (\text{final velocity} \times \text{time})}{\text{time}}$

$= \dfrac{2 \times 300 \text{ m} - (18 \text{ ms}^{-1} \times 20 \text{ s})}{20 \text{ s}}$

$= 12$ ms^{-1}

2.11) A dog has a displacement of 20 metres from its starting position. If it had an initial velocity of 1 ms^{-1} and a final velocity of 4 ms^{-1}, calculate the time for which it was travelling.

You have displacement (s), initial velocity (u) and final velocity (v). You need time (t) so use equation 4.

$$\frac{2s}{(u+v)t}$$

As time (t) is on the bottom, you can just cover it over to find the equation you need.

time $= \dfrac{2 \times \text{displacement}}{\text{initial velocity} + \text{final velocity}}$

$= \dfrac{2 \times 20 \text{ m}}{1 \text{ ms}^{-1} + 4 \text{ ms}^{-1}}$

$= 8$ s

$$t = \frac{2s}{u+v}$$

b) What are you assuming?

You are assuming that the acceleration is constant.

2.12) An eagle travels in a straight line through a distance of 4 km in a time of 10 minutes. If its initial velocity was 10 ms^{-1} and assuming constant acceleration, what is the value of its final velocity?

Convert km into metres: 2 km = 2,000 m
Convert minutes into seconds: 10 min = 600 s

You have displacement (s), time (t) and initial velocity.
You need final velocity (v) so use equation 4.

To separate out the final velocity (v) from this equation, you need to multiply out the time first.

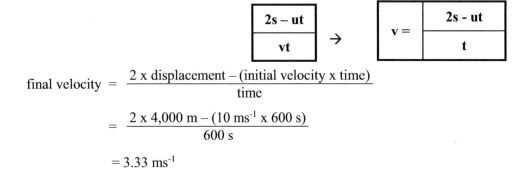

Rule 1: Now you can put the 'ut' on the other line and cover over 'v'.

final velocity = $\dfrac{2 \times \text{displacement} - (\text{initial velocity} \times \text{time})}{\text{time}}$

$= \dfrac{2 \times 4{,}000 \text{ m} - (10 \text{ ms}^{-1} \times 600 \text{ s})}{600 \text{ s}}$

$= 3.33 \text{ ms}^{-1}$

2.13) A small electric toy has an initial velocity of 0.1 ms^{-1}. It is accelerating down a ramp towards a jump. The toy accelerates at a rate of 0.05 ms^{-2} for a period of 8 seconds when it reaches the jump. Calculate the final velocity of the car as it hits the jump.

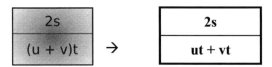

You have initial velocity (u), acceleration (a) and time (t).
You need final velocity (v) so use equation 1.

final velocity = initial velocity + (acceleration x time)
$= 0.1 \text{ ms}^{-1} + (0.05 \text{ ms}^{-2} \times 8 \text{ s})$
$= 0.5 \text{ ms}^{-1}$

2.14) A moped is travelling along a smooth section of road. The moped accelerates at a rate of 1.2 ms^{-2} for a time of 6 seconds. It reaches a final velocity of 12.5 ms^{-1}. What was the initial velocity of the moped?

You have acceleration (a), time (t) and final velocity (v).
You need initial velocity (u) so use equation 1.

Rule 1: You just need to move 'at' to the other line by changing its sign.

initial velocity = final velocity − (acceleration x time)
= 12.5 ms^{-1} − (1.2 ms^{-2} x 6 s)
= 5.3 ms^{-1}

2.15) A cart is attached to the back of a small 4-wheel drive ATV. It accelerates at a rate of 1.3 ms^{-2}, changing its velocity from 1.6 ms^{-1} to 4.8 ms^{-1}. Calculate the time for which the cart is accelerated.

You have acceleration (a), initial velocity (u) and final velocity (v).
You need time (t) so use equation 1.

Rule 1: You are looking for time (t) so move the 'u' to the other line by changing the sign and then cover up the 't'.

time = $\dfrac{\text{final velocity} - \text{initial velocity}}{\text{acceleration}}$

= $\dfrac{4.8 \text{ ms}^{-1} - 1.6 \text{ ms}^{-1}}{1.3 \text{ ms}^{-2}}$

= 2.5 s

$t = \dfrac{v - u}{a}$

2.16) Calculate the acceleration required for a hawk to reach a final velocity of 20 ms^{-1} from an initial velocity of 4 ms^{-1} in a time of 3 seconds.

You have final velocity (v), initial velocity (u) and time (t).
You need acceleration (a) so use equation 1.

Rule 1: Place the 'u' on the other line by changing its sign and then cover up the 'a' to find your equation.

acceleration = $\dfrac{\text{final velocity} - \text{initial velocity}}{\text{time}}$

= $\dfrac{20 \text{ ms}^{-1} - 4 \text{ ms}^{-1}}{3 \text{ s}}$

= 5.3 ms^{-2}

$a = \dfrac{v - u}{t}$

Bonus Questions 2.1: Equations of Motion and More Complex Movement

2.1.1) Calculate the acceleration of a rabbit with an initial velocity of 1.5 ms^{-1} and a final velocity of 6 ms^{-1}. It increases its speed over a distance of 4 m.

2.1.2) A bus is accelerating at a constant rate of 2 ms^{-1}. If it starts from rest, calculate the distance that the bus covers when it is accelerating between the following velocities.
   a) 0 ms^{-1} and 4 ms^{-1}
   b) 4 ms^{-1} and 8 ms^{-1}
   c) 8 ms^{-1} and 12 ms^{-1}

2.1.3) A skateboarder is travelling downhill with an acceleration of 2 ms^{-2}. If they reach a final velocity of 15 ms^{-1} over a distance of 50 m, what was their initial velocity?

2.1.4) A truck has an initial velocity of 14 ms^{-1} and accelerates over a distance of 100 m at a rate of 1 ms^{-2}. Calculate the final velocity of the truck.

2.1.5) A glass marble rolls down a smooth slope. It travels through a displacement of 1.8 m in a time of 2 seconds. If it had an acceleration of 0.5 ms^{-2} what was the value of its initial velocity?

2.1.6) An object falls from a height of 10 m to hit the ground in a time of 1.43 s. Assuming it fell from rest, calculate its acceleration.

2.1.7) For the previous question, what would the acceleration of the object be if it had an initial velocity of:
   a) 3 ms^{-1} downward towards the ground?
   b) 3 ms^{-1} upward away from the ground?

2.1.8) A toy runs with a series of little feet operated by a wheel. It is placed at the top of a small slope and accelerates from a stationary position down to the base of the slope which is 84 cm away. If the toy experiences an acceleration of 0.05 ms^{-2} calculate the time required for the toy to travel to the base of the slope.

2.1.9) A boat captain has turned on his engines. The boat is initially travelling down a river with a velocity of 1.2 ms^{-1}. The boat undergoes an acceleration of 0.8 ms^{-2} for a period of 7 seconds. If it was opposite a person on the bank of the river at the beginning of this acceleration, what is the value of its final displacement from that same person after this time?

2.1.10) A bicyclist is travelling at an initial rate of 3 ms^{-1}. He accelerates at a constant rate to a final velocity of 10 ms^{-1} over a time period of 7 seconds. What is his displacement from his original position after this time?

2.1.11) A dolphin accelerates from a velocity of 5 ms^{-1} to 12 ms^{-1} in a time of 8 seconds. By how much does the displacement of the dolphin change during this period?

2.1.12) A walker accelerates at a constant rate to a velocity of 3 ms^{-1}. If they cover a distance of 100 metres in a time of 50 seconds, what was their original velocity?

2.1.13) One of the most successful early steam trains was the rocket. It had a maximum speed that would allow it to cover a distance of 10 km in a time of 800 seconds. Assuming that the locomotive was initially stationary, the acceleration was constant and the maximum speed was reached at 10 km, calculate the maximum speed of the steam engine.

2.1.14) A cart is accelerating down a long hill. The cart has an initial speed of 1.5 ms^{-1} and it covers a distance of 1 km in a time of 300 seconds.
   a) What is the maximum speed of the cart?
   b) What are you assuming?

2.1.15) A throwing aid allows dog owners to throw tennis balls long distances with ease. A tennis ball is placed on the end of the throwing aid and the ball is accelerated at a rate of 60 ms^{-2} for a period of 0.3 seconds. If it had an initial velocity of 1 ms^{-1}, calculate the final velocity of the tennis ball.

2.1.16) An emu is running across a field. It accelerates at a rate of 3 ms^{-2} for a period of 3 seconds before it finally reaches a speed of 14 ms^{-1}. What was the initial speed of the emu?

2.1.17) A horse accelerates at a constant rate, changing its speed from 4.5 ms^{-1} to 14 ms^{-1} in a time of 6 seconds. What is the acceleration of the horse?

2.1.18) A toy submarine accelerates to a velocity of 0.68 ms^{-1} from an initial velocity of 0.12 ms^{-1}. If this change of velocity occurred over a time of 12 seconds, calculate the acceleration of the toy submarine.

Bonus Questions 2.2: Equations of Motion and More Complex Movement

2.2.1) A small dog has an initial velocity of 0.8 ms^{-1} and a final velocity of 4.8 ms^{-1}. It increases its speed over a distance of 6 m. Calculate the acceleration of the dog.

2.2.2) A motorbike on a racetrack is accelerating at a constant rate of 5 ms^{-2}. Calculate the distance that the motorbike covers when it is accelerating between the following velocities.
   a) 0 ms^{-1} and 5 ms^{-1}
   b) 10 ms^{-1} and 15 ms^{-1}
   c) 20 ms^{-1} and 25 ms^{-1}

2.2.3) A boulder is being removed for safety by rolling it down the side of a steep slope away from a road. It has an acceleration of 3 ms^{-2}. It has a final velocity of 18 ms^{-1} and it achieves this after travelling a distance of 45 m. What was the initial velocity of the boulder once it was pushed?

2.2.4) A small van has an initial velocity of 14 ms^{-1} and accelerates at a rate of 0.8 ms^{-2} over a distance of 78 m. Calculate the final velocity of the van.

2.2.5) A tennis ball rolls down a hill. It travels through a displacement of 30 m in a time of 12 seconds. If it had an acceleration of 0.3 ms^{-2}, what was the value of its initial velocity?

2.2.6) A stone falls from a height of 50 m to hit the ground in a time of 3.19 s. If it fell from rest, calculate the value of its acceleration.

2.2.7) For the previous question, what would the acceleration of the object be if it had an initial velocity of:
   a) 10 ms^{-1} downward towards the ground?
   b) 10 ms^{-1} upward away from the ground?

2.2.8) A small toy dog moves by means of a wheel that it stands on. It is wound up and then released from a stationary position along a flat surface. It travels a distance of 2.34 m and experiences an acceleration of 0.12 ms^{-2}. Calculate the time required for the toy to travel this distance.

2.2.9) A barge is being pulled by a motor connected to a pulley. It is initially travelling along a stretch of water with a velocity of 0.1 ms^{-1}. The boat undergoes an acceleration of 0.12 ms^{-2} for a period of 8 seconds. If it was next to a mooring point at the start, what was the barge's displacement from its original position in this time?

2.2.10) A skateboarder is travelling at an initial rate of 0.1 ms^{-1}. She accelerates at a constant rate to a final velocity of 5 ms^{-1} over a time period of 7 seconds. What is her displacement from her original position after this time?

2.2.11) A racehorse accelerates from a velocity of 2 ms^{-1} to 11 ms^{-1} in a time of 5 seconds. By how much does the displacement of the racehorse change during this period?

2.2.12) A jogger accelerates at a constant rate to a velocity of 5 ms^{-1}. If they cover a distance of 150 m in a time of 40 s, what was their original velocity?

2.2.13) Over a hundred years ago a fast paddle steamer could reach its maximum speed if it accelerated from stationary at a constant rate for a period of 93 seconds. It would cover a distance of 432 meters in this time. Calculate the value of its maximum speed.

2.2.14) A car is near the top of a hill with a flat battery. It is rolling down the hill to try to gain enough speed so that the driver can start it. The car has an initial speed of 0.6 ms^{-1} and it covers a distance of 45 m in a time of 40 seconds.
   a) What is the maximum speed of the car?
   b) What are you assuming?

2.2.15) An experimental trebuchet has been built by some university students to test their theories. It is designed to throw basketballs long distances in a field. A basketball is placed on the end of the trebuchet and is accelerated at a rate of 80 ms^{-2} for a period of 0.4 seconds. If it had an initial velocity of 1.5 ms^{-1}, calculate the final velocity of the basketball.

2.2.16) A dog is running across a field. It accelerates at a rate of 2.5 ms^{-2} for a period of 3.2 seconds before finally it reaches a speed of 12 ms^{-1}. What was the initial speed of the dog?

2.2.17) A zebra undergoes constant acceleration and changes its speed from 1.2 ms^{-1} to 12 ms^{-1} in a time of 4 seconds. What is the acceleration of the zebra?

2.2.18) A duck swimming on the surface of a lake accelerates to a velocity of 3.2 ms^{-1} from an initial velocity of 0.3 ms^{-1}. If this change of velocity occurred over a time of 4 seconds calculate the acceleration of the duck.

3. Momentum, Force, Moment and Mass

Vikings Like Dancing

Make Vikings **m**ove.
Make Vikings **d**ance.
Must **f**ind **d**ancers
for **e**very **s**ong.

Make **V**ikings **m**ove.	m* v x m	Momentum = Velocity × Mass
Make **V**ikings **d**ance.	m v x d	Mass = Volume × Density
Must **f**ind **d**ancers	m f x d	Moment = Force × Distance
for **e**very **s**ong.	f e x s (also f = kx)	Force = Extension × Spring Constant

*Momentum is often written using the lower case p.

When answering questions in exams I recommend writing the full equations down with words to make sure that you get the most marks available. Otherwise pay close attention to the letters that are used by your teacher and ensure you use these.

In exams lots of questions on this topic might ask you how to find the volume of an object. Make sure that you know how to do this! Near the end of this book is a section that gives you the calculations for finding area and volume for the main shapes that you may come across.

Your Favourite Food

And remember…

Total momentum is like your favourite food. It just **doesn't change**.

Total momentum before = Total momentum after

| Total momentum before | = | Total momentum after |

Examples: Momentum, Force, Moment and Mass

Example 1) An object has a mass of 10 kg and a volume of 1 m³. What is its density?

density = $\dfrac{\text{mass}}{\text{volume}}$

= $\dfrac{10 \text{ kg}}{1 \text{ m}^3}$

= 10 kgm⁻³

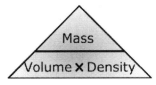

Example 2) An object has a density of 3 kgm⁻³. It has a volume of 0.8 m³. What is its mass?

mass = volume x density
= 0.8 m³ x 3 kgm⁻³
= 2.4 kg

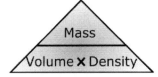

Example 3) A spring is extended 20 cm by a force of 10 N. What is the spring constant of the spring?

Distance is measured in metres 20 cm = 0.2 m

spring constant = $\dfrac{\text{force}}{\text{extension}}$

= $\dfrac{10 \text{ N}}{0.2 \text{ m}}$

= 50 Nm⁻¹

Example 4) A spring of spring constant 50 Nm⁻¹ has a force of 10 N applied to it. Another spring of spring constant 30 Nm⁻¹ has a force of 8 N applied to it. Which spring has the largest extension?

extension = $\dfrac{\text{force}}{\text{spring constant}}$

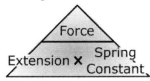

Spring 1

extension = $\dfrac{10 \text{ N}}{50 \text{ Nm}^{-1}}$

= 0.2 m

Spring 2

extension = $\dfrac{8 \text{ N}}{30 \text{ Nm}^{-1}}$

= 0.27 m

Spring 2 has the largest extension.

Example 5) What makes an object stable? (This does not link to any of the equations in this book but it is very important in the principle of stability).

An object is stable when it has a low centre of gravity and a wide base.

Example 6) A child of mass 30 kg sits 2 metres away from the pivot of a see-saw. How far must the child's older sister sit from the pivot on the other side to balance the see-saw? She has a mass of 40 kg.

weight (this is the force) = mass x gravity
moment = force x distance
moment 1 = 30 kg x 10 ms^{-2} x 2 m
 = 600 Nm

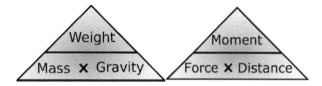

moment 2 = 40 kg x 10 ms^{-2} x distance 2

For the see-saw to balance the moment at either side must be equal.

600 Nm = 400 N x distance 2

distance 2 = 600 Nm / 400 N
 = 1.5 m

Example 7) A mass of 10 kg is moving at a velocity of 3ms^{-1}. It collides with a 40 kg mass which is stationary. The masses stick together as they move on.
What is the final velocity of the objects?

momentum = velocity x mass

With all of these questions it is essential to do the following:

1. Calculate the value of the initial momentum and label it momentum before or initial momentum.

2. Write momentum before = momentum after.

| Total momentum before | = | Total momentum after |

If these two suggestions are followed you will generally have at least 50% of the marks for the momentum part of the collision questions. Alternately, if you do not write them, you could lose up to 50% of the marks even if everything is 100% correct.

momentum before = m$_1$v$_1$ + m$_2$v$_2$
 = (10 kg x 3 ms^{-1}) + (40 kg x 0 ms^{-1})
 = 30 kgms^{-1} or 30 Ns

momentum before = momentum after

So: $v_f = \dfrac{m_1 v_1 + m_2 v_2}{m_f}$

The final mass is just the two separate masses added together = 40 kg + 10 kg
= 50 kg

$$v_f = \frac{(10 \text{ kg} \times 3 \text{ ms}^{-1}) + (40 \text{ kg} \times 0 \text{ ms}^{-1})}{50 \text{ kg}}$$

$$= \frac{30 \text{ kgms}^{-1}}{50 \text{ kg}}$$

$$= 0.6 \text{ ms}^{-1}$$

This value for the velocity is positive. This means that it is travelling in the same direction as the 20 kg mass was.

Questions: Momentum, Force, Moment and Mass

3.1) An object of mass 2 kg travels with a velocity of 20 ms^{-1} towards a stationary target of mass 30 kg. They collide and stick together. What is the final velocity of the objects?

3.2) 30 bags of sugar each of mass 1 kg are contained in a box on one side of a pivot balanced by a mass of 50 kg on the other side. 15 bags of sugar are removed. What mass should now be on the other side to maintain the balance? Assume that the box has no mass.

3.3) A cube of length 3 m has a mass of 2,000 kg. What is its density?

3.4) A 30,000 kg truck travelling at 10 ms^{-1} collides with a car of mass 1,000 kg travelling at -10 ms^{-1}. The truck and car stick together. What is their final velocity?

3.5) A spring of spring constant 20 Nm^{-1} is compressed by 25 cm. How much force is being applied to the spring?

3.6) I have 18,000 kg of a material of density 2,000 kgm^{-3}. What is its volume?

3.7) Archimedes is famous for many things however one of his greatest achievements was being able to prove that the king's sacred golden crown was not made purely from gold. The man who made the crown had stolen some of the gold and mixed in some cheap material instead so it looked to be the same amount.

Using the equations that you have already learned from the book how could you prove the crown was not made of pure gold?
The density of gold is: 19.32 g/cm^3.

3.8) A ship is floating on the surface of the water. If it displaces a volume of 6.5 m^3 of water, then what is the mass of the ship?
The density of water is 1,000 kgm^{-3}.

3.9) A fisherman wants to measure the mass of a fish he has caught. All he has is a spring of spring constant 300 Nm^{-1}. Using the spring he lifts the fish and measures the extension of the spring. The extension of the spring is 30 cm. What is the mass of the fish?

3.10) Using values from the graph, which spring has the greatest spring constant? Which of the springs reaches its limit of proportionality first?

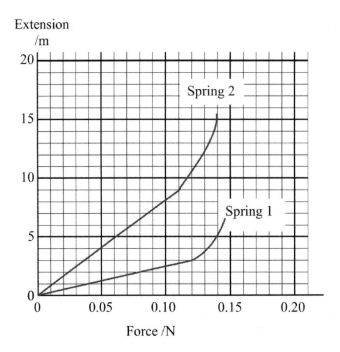

3.11) An object of mass 0.3 kg is placed on a lever 3 metres from a pivot. What is the moment that the object develops? This lever is prevented from closing by a small metal nut that is located 5 centimetres from the pivot. What is the force felt by the nut?

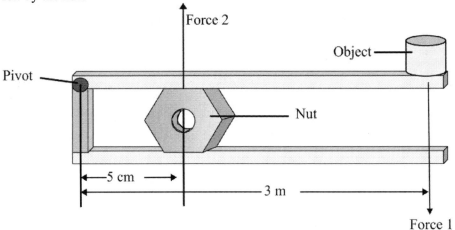

3.12) Explain how a lever can be used to increase a force.

Answers: Momentum, Force, Moment and Mass

3.1) An object of mass 2 kg travels with a velocity of 20 ms^{-1} towards a stationary target of mass 30 kg. They collide and stick together. What is the final velocity of the objects?

total momentum before = total momentum after

momentum = velocity x mass

Our first step is to calculate the initial momentum before the objects collide.

$$\text{momentum before} = m_1v_1 + m_2v_2$$
$$= 2 \text{ kg} \times 20 \text{ ms}^{-1} + 30 \text{ kg} \times 0 \text{ ms}^{-1}$$
$$= 40 \text{ kgms}^{-1} \text{ or } 40 \text{ Ns}$$

momentum before = momentum after

So: $v_f = \dfrac{m_1v_1 + m_2v_2}{m_f}$

The final mass is the two separate masses together = 2 kg + 30 kg = 32 kg

$$v_f = \frac{(2 \text{ kg} \times 20 \text{ ms}^{-1}) + (30 \text{ kg} \times 0 \text{ ms}^{-1})}{32 \text{ kg}}$$

$$= \frac{40 \text{ kgms}^{-1}}{32 \text{ kg}}$$

$$= 1.25 \text{ ms}^{-1}$$

This value for the velocity is positive. This means that it is travelling in the same direction as the 20 kg mass was.

3.2) 30 bags of sugar each of mass 1 kg are contained in a box on one side of a pivot balanced by a mass of 50 kg on the other side. 15 bags of sugar are removed. What mass should now be on the other side to maintain the balance? Assume that the box has no mass.

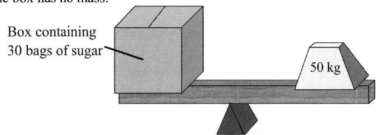

If the pivot is balanced, then the moments are equal

moment 1 = moment 2
force 1 x distance 1 = force 2 x distance 2

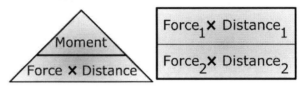

The force comes from the weight (weight = gravity x mass) and we are assuming that when we replace the second mass we will put it at the same distance from the pivot.

So our initial equation is:

mass of 30 bags of sugar x gravity x distance 1 from pivot = 50 kg x gravity x distance 2 from pivot

We can divide both sides by gravity just so it looks a little neater.
mass of 30 bags of sugar x distance 1 = 50 kg x distance 2

Divide both sides by 30 to find how much is needed to balance a single bag of sugar.
mass of 1 bag of sugar x distance 1 = 1.67 kg x distance 2

I want to balance 15 bags of sugar so I multiply both sides by 15.
mass of 15 bags of sugar x distance 1 = 25 kg x distance 2

I will need to replace the 50 kg mass by a 25 kg mass.

3.3) A cube of length 3 m has a mass of 2,000 kg. What is its density?

volume of cube = length x depth x height
 = 3m x 3m x 3m
 = 27 m^3

$$\text{density} = \frac{\text{mass}}{\text{volume}}$$
$$= \frac{2{,}000 \text{ kg}}{27 \text{ m}^3}$$
$$= 74 \text{ kg}/\text{m}^3$$

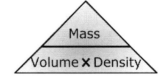

3.4) A 30,000 kg truck travelling at 10 ms⁻¹ collides with a car of mass 1,000 kg travelling at -10 ms⁻¹. The truck and car stick together. What is their final velocity?

momentum before = momentum after
$= m_1v_1 + m_2v_2$
momentum after = m_fv_f

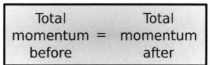

We want to find the final velocity v_f

So: $v_f = \dfrac{m_1v_1 + m_2v_2}{m_f}$

The final mass is just the two separate masses added together = 30,000 kg + 1,000 kg
= 31,000 kg.

$v_f = \dfrac{(30,000 \text{ kg} \times 10 \text{ ms}^{-1}) + (1,000 \text{ kg} \times -10 \text{ ms}^{-1})}{31,000 \text{ kg}}$

$= \dfrac{290,000 \text{ kgms}^{-1}}{31,000 \text{ kg}}$

$= 9.35 \text{ ms}^{-1}$

This value for the velocity is positive. The car and truck are both travelling at 9.35 ms⁻¹ in the original direction of the truck.

3.5) A spring of spring constant 20 Nm⁻¹ is compressed by 25 cm. How much force is being applied to the spring?

force = extension x spring constant

Extension, like all distances, is measured in metres
25 cm = 0.25 metres

force = 0.25 m x 20 Nm⁻¹
= 5 N

3.6) I have 18,000 kg of a material of density 2,000 kgm⁻³. What is its volume?

volume = $\dfrac{\text{mass}}{\text{density}}$

$= \dfrac{18,000 \text{ kg}}{2,000 \text{ kgm}^{-3}}$

$= 9 \text{ m}^3$

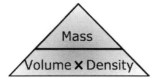

3.7) Archimedes is famous for many things however one of his greatest achievements was being able to prove that the king's sacred golden crown was not made purely from gold. The man who made the crown had stolen some of the gold and mixed in some cheap material instead so it looked to be the same amount.

Using the equations that you have already learned from the book how could you prove the crown was not made of pure gold?
The density of gold is: 19.32 g/cm³.

Answer: The only way to prove that it isn't gold without cutting pieces out of it is to prove that its density isn't 19.32 g/cm³.

All you need to calculate the density is the mass and the volume of the object. The volume can be found by placing the crown in water and examining how much the water level raises. The difference in volume (volume of water with the crown - volume of water without the crown) will give you volume of the crown.

You can weigh the crown to find its mass. Placing the numbers into the equation for density

$$\text{density} = \frac{\text{mass}}{\text{volume}}$$

will give you the density of the crown. If this is less than the density of gold, then the crown is not made of pure gold. This is the answer that is expected on tests.

It is likely that Archimedes used the following setup instead.

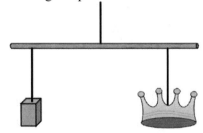

The position of gold the on the left hand side is adjusted so that the moment created by the weight of the gold on the left hand side (the anti-clockwise moment) is exactly equal to the moment created by the weight of the crown (the clockwise moment).

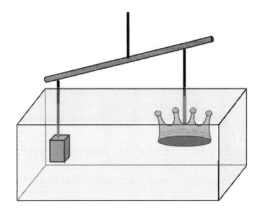

The gold and the crown were then lowered slowly into the water. If the crown was made of pure gold, then moment from both weights will still be equal when they are in the water. If the moments are different the balance will not remain level. This means that they have different densities and the crown is not pure gold.

Using this method Archimedes would have been able to prove that the crown was not pure gold without ever damaging it or performing a single calculation!

3.8) A ship is floating on the surface of the water. If it displaces a volume of 6.5 m³ of water, then what is the mass of the ship?
The density of water is 1,000 kgm^{-3}.

For an object to float it must displace the equivalent of its own mass in liquid.
This is called the Archimedes principle.

mass = volume x density
mass of water displaced = 6.5 m³ x 1,000 kgm^{-3}
                                  = 6,500 kg
                                  = mass of ship
mass of ship = 6,500 kg.

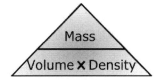

3.9) A fisherman wants to measure the mass of a fish he has caught. All he has is a spring of spring constant 300 Nm^{-1}. Using the spring he lifts the fish and measures the extension of the spring. The extension of the spring is 30 cm. What is the mass of the fish?

Our first step is to find the force needed to extend the spring by this amount.

force = extension x spring constant
extension is measured in metres.
extension = 0.3 metres
force = 0.3 m x 300 Nm^{-1}
      = 90 N

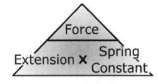

This force comes from the weight of the fish and we wish to calculate the mass.

$$mass = \frac{weight}{gravity}$$

gravity = 10 ms⁻²

$$mass = \frac{90 \text{ N}}{10 \text{ ms}^{-2}}$$

mass of fish = 9 kg

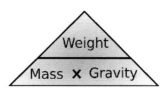

3.10) Using values from the graph, which spring has the greatest spring constant? Which of the springs reaches limit of proportionality first?

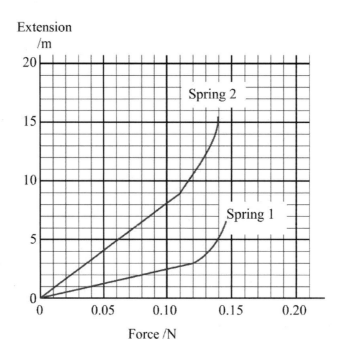

$$spring\ constant = \frac{force}{extension}$$

Spring 1
k = 0.12 N / 0.03 m
= 4 Nm⁻¹

Spring 2
k = 0.11 N / 0.09 m
= 1.22 Nm⁻¹

Examining the graph, we can see that Spring 2 has reached the limit of its proportional extension (where the straight line section of the graph ends) at 0.11 newtons. Spring 1 reaches its limit of proportionality at 0.12 newtons. Spring 2 will reach its elastic limit first.

3.11) An object of mass 0.3 kg is placed on a lever 3 metres from a pivot. What is the moment that the object develops? This lever is prevented from closing by a small metal nut that is located 5 centimetres from the pivot. What is the force felt by the nut?

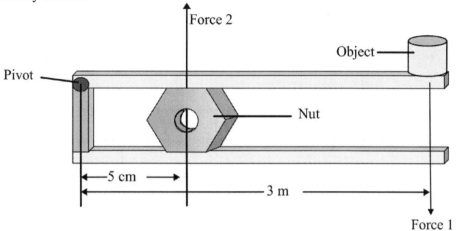

The force acting to create the moment comes from the weight of the object.

weight = mass x gravity
 = 0.3 kg 10 ms^{-2}
 = 3 N

moment = force x distance
 = 3 N x 3 metres
 = 9 Nm

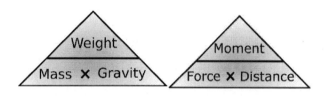

The nut, in order to avoid breaking, will need to create the same force but in the opposite direction.

moment = force x distance = - 9 Nm

distance = 0.05 m

force = $\dfrac{\text{moment}}{\text{distance}}$

 = $\dfrac{-9 \text{ Nm}}{0.05 \text{ m}}$

 = -180 N

The negative here tells us that the force is in the opposite direction to the original force. This means that the force produced by the nut is then:

force = 180 N upwards

3.12) Explain how a lever can be used to increase a force.

The lever is a force multiplier. It works because it is combined with a pivot or fulcrum.

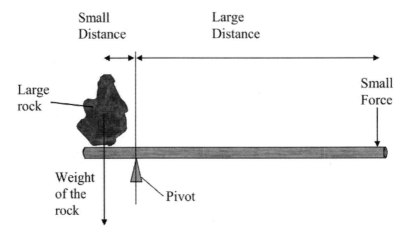

Something happens:
We apply a force to the longer end of the lever.

Which has an effect:
This will create a moment due to the distance from the pivot.

Which means:
In order to prevent the lever from turning there will need to be an equal moment acting in the opposite direction that is provided by a force at the shorter end of the lever.

Therefore:
As moment = force x distance and the distance to the pivot is smaller the force must be larger.

Final answer:
We apply a force to the longer end of the lever. This will create a moment due to the distance from the pivot. In order to prevent the lever from turning there will need to be an equal moment acting in the opposite direction that is provided by a force at the shorter end of the lever. As moment = force x distance and the distance to the pivot is smaller the force must be larger.

Bonus Questions 3.1: Momentum, Force, Moment and Mass

3.1.1) Calculate the moment that is created by hanging a shop sign of mass 4 kg a distance of 50 cm from a pivot.

3.1.2) A cat of mass 3 kg is sitting 1.2 metres from a pivot.
   a) Calculate the moment that this generates.
   b) After a little while the cat moves to a new seating position 30 cm from the pivot. Calculate the new moment that this creates.
   c) Calculate the difference in moment that the cat created between the first and second seating position.

3.1.3) A beam has a pivot at its centre and its weight is equally distributed so that it will normally balance. Two students are playing a game where they each take an object of different weights and guess where on the beam to place them. Then they both release their object and test whether the beam will balance. The first student has an apple of weight 1.12 N and they place it a distance of 20 cm from the pivot so that it produces a clockwise moment. The second student places an eraser of weight 0.4 N a distance of 56 cm from the pivot so that it produces an anticlockwise moment. Does the beam balance?

3.1.4) A metre ruler has a pivot in the centre at the 50 cm mark. A 5 kg mass is suspended from the ruler at the 80 cm mark.
   a) Calculate the moment that this creates about the pivot.
   b) At what location would an 8 kg mass need to be suspended in order for the metre rule to balance?

3.1.5) A force of 1,000 N is needed to open a manhole cover. The manhole is being prized off with a crowbar that has a pivot 1 cm from the contact point with the cover. I use a force of 30 N at distance of 34 cm from the pivot.
   a) Can I lift the manhole cover?
   b) What is the minimum force that I need to use at this point to open the manhole cover?

3.1.6) A box is made up with a volume of 1.2 m^3. It is filled completely with a material of density 150 kgm^{-3}. Calculate the maximum mass that the box must be able to hold.

3.1.7) A crate is able to hold a weight of 5,600 newtons. It is loaded with a material of density 2,440 kgm^{-3}. Calculate the maximum volume of the material that the crate can hold.

3.1.8) A rectangular container has sides of length 20 cm wide, 60 cm long and 15 cm high. It is filled completely with a liquid. When the liquid is weighed it has a weight of 2,780 newtons. Calculate the density of the liquid.

3.1.9) An object has a mass of 50 kg and a volume of 4,000 cm^3. Calculate the density of the material that the object is made from.
   a) Write this in both kgm^{-3} and gcm^{-3}.
   b) What mass of material would I have if I had 300 cm^3 of the material?

3.1.10) Calculate the volume of material needed to create a 30 kg weight if the material being used has a density of 400 kgm^{-3}.

3.1.11) Water has a density of 1,000 kgm^{-3}. Calculate this density in gcm^{-3}.

3.1.12) A car has a mass of 1,760 kg travelling with a velocity of 20 ms^{-1}.
   a) Calculate the momentum of the car.
   b) Then calculate the car's kinetic energy.
   c) Calculate the momentum of a small truck of mass 3,520 kg travelling with a velocity of 10 ms^{-1}.
   d) Calculate the KE of the small truck and compare the values of momentum and KE for both the truck and the car.

3.1.13) A man of mass 75 kg jumps onto a small boat of mass 50 kg with a speed of 3 ms^{-1} and stops. Calculate the initial speed of the boat and man after he jumps in.

3.1.14) The International Space Station has a mass of 375,000 kg and is travelling at a speed of 27,600 km/h. Calculate its momentum.

3.1.15) Calculate the momentum of a truck of mass 5,000 kg travelling at a speed of 0.3 ms^{-1}. The truck drives onto a ferry of mass 30,000 kg. It collides gently with the walls of the ferry and stops. Calculate the initial speed of the truck and ferry after the collision.

3.1.16) A piece of ice of mass 30 g is travelling at a speed of 2 ms^{-1} along a frictionless surface. It collides with another block of ice and sticks to it. If the mass of the stationary ice is 200 g, calculate the final speed of the ice immediately after the collision.

3.1.17) A spring is extended by a distance of 15 cm when it is holding a weight of 15 N.
   a) Calculate the value of its spring constant.
   b) What would the value of its extension be if the original weight of 15 N was removed and a mass of 1 kg was added and the spring had to support its entire weight?

3.1.18) A spring is extended by a force of 40 newtons. It has a spring constant of 200 Nm^{-1}.
   a) Calculate the value of its extension.
   b) If the original length of the spring is 12 cm how long will it be when it is being extended by the force.

3.1.19) Two identical springs each of spring constant 120 Nm^{-1} are attached end to end and the top spring is suspended from a nail in a wall. A force of 30 N is applied to the base of the bottom spring how far will the end of the bottom spring move because of the force applied?

3.1.20) A spring of spring constant 24 Nm^{-1} is extended a distance of 30 cm.
   a) What force is required?
   b) The force comes from the weight of a fish which is being weighed on a spring balance scale at a fish market. What is the mass of the fish?

3.1.21) A spring has an initial length of 20 cm.
   a) Calculate the spring constant if a force of 50 N is required to change its length to 54 cm.
   b) The same spring is then extended to a final length of 60 cm. What was the total final force being applied to the spring?

Bonus Questions 3.2: Momentum, Force, Moment and Mass

3.2.1) What is the value of the moment that is created by hanging a basket of flowers of mass 3 kg a distance of 30 cm from a pivot?

3.2.2) A box of tools of mass 5.4 kg is placed 45 cm from a pivot.
   a) Calculate the moment that this generates.
   b) After a little while the box is moved to a new position 1.8 m from the pivot. Calculate the new moment that this creates.
   c) Calculate the difference in moment that the box created between the first and second position.

3.2.3) A balanced beam is pivoted at its centre. A book of weight 2.6 N is placed a distance of 28 cm from the pivot so that it produces a clockwise moment. A 3.1 N paperweight is placed at a distance of 23 cm from the pivot so that it produces an anticlockwise moment. Does the beam balance?

3.2.4) A meter ruler is pivoted about its centre at the 50 cm mark. A 3 kg mass is suspended from the ruler at the 65 cm mark.
   a) Calculate the moment that this creates about the pivot.
   b) What location would a 7 kg mass need to be suspended at in order for the meter rule to balance?

3.2.5) A force of 300 N is needed to open a can of paint. The paint can is being prized open using a flat head screwdriver that is being pivoted on the edge of the can a distance of 5 mm from the edge of the lid. I use a force of 20 N at a distance of 15 cm from the pivot.
   a) What is the moment that is generated by the force applied to the screwdriver?
   b) What is the minimum force that I need to use to open the can of paint?

3.2.6) A large chest has a volume of 3.6 m^3. It is filled with a material of density 340 kgm^{-3}. Calculate the minimum mass that the box must be able to hold.

3.2.7) A small concrete base is able to hold a weight of 38,000 newtons before cracking. It is loaded with a material of density 3,500 kgm^{-3}. Calculate the maximum volume of the material that can be placed on the concrete base.

3.2.8) A cube has sides of length 20 cm. It is filled completely with a liquid. When the liquid is weighed it has a weight of 122 newtons. Calculate the density of the liquid.

3.2.9) A small ceramic box has a mass of 30 kg and a volume of 3,500 cm^3. Calculate the density of the material that the object is made from.
   a) Write this in both kgm^{-3} and gcm^{-3}.
   b) What mass of material would I have if I had 300 cm^3 of the material?

3.2.10) Calculate the volume of material needed to create a 50 kg weight if the material being used has a density of 3,000 kgm^{-3}.

3.2.11) Mercury has a density of 13,690 kgm^{-3}. Calculate this density in gcm^{-3}.

3.2.12) A small van of mass 2,350 kg is travelling with a velocity of 24 ms^{-1}.
   a) Calculate the momentum of the van.
   b) Calculate the momentum of a truck of mass 2,850 kg travelling in the opposite direction with a speed of 14 ms^{-1}.
   c) The small van collides with the truck and they stick together. Calculate the final velocity of the truck and van.

3.2.13) A man of mass 65 kg steps onto a stationary canoe of mass 25 kg with a speed of 3 ms^{-1} and stops walking. Calculate the combined velocity of the canoe and man after he steps in.

3.2.14) The Hubble space telescope has a mass of 11,110 kg and is travelling at a speed of 8 kms^{-1}. Calculate its momentum.

3.2.15) Calculate the momentum of a bus of mass 4,800 kg travelling at a speed of 0.6 ms^{-1}. The bus drives onto a 120,000 kg stationary boat and stops. Calculate the final speed of the bus and boat after the bus stops.

3.2.16) A 60 kg ice skater is travelling at a speed of 3 ms^{-1} along a frictionless surface. They pick up a stationary bag of mass 12 kg from the ice. Calculate the final velocity of the ice skater immediately after they pick up the bag.

3.2.17) A spring is extended by a distance of 30 cm when it is holding a weight of 12 N.
   a) What is the value of the spring constant of the spring?
   b) What would the value of its extension be if the original weight of 12 N was removed and a mass of 0.15 kg was added and the spring had to support its entire weight?

3.2.18) A spring is supporting a weight of 23 newtons. It has a spring constant of 100 Nm^{-1}.
   a) Calculate the value of its extension.
   b) The spring has a natural length of 2.6 cm. Calculate the total length of the spring while it is supporting the weight.

3.2.19) Two identical springs, each of spring constant 80 Nm^{-1} are attached end to end and the top spring is suspended from a clamp stand. A force of 15 N is applied to the base of the bottom spring. How far will the end of the bottom spring move because of the force applied?

3.2.20) A student is performing an experiment. She measures the length of a spring that is holding a cup. She then adds beads to the cup and measures the extension of the spring. The spring has a spring constant 30 Nm^{-1} and an extension of 12 cm.
   a) How much force does the student need to apply to the spring to extend it this amount?
   b) Each bead has a mass of 1g. How many beads are in the cup?

3.2.21) A spring has an initial length of 15 cm.
   a) A force of 30 N is applied to the spring. The total length of the spring becomes 45 cm. Calculate the spring constant of the spring.
   b) The same spring is then extended to a final length of 60 cm. What was the total final force being applied to the spring?

4. Kinetic Energy, Potential Energy and Energy Stored in Mass

The Animals' Lunchtime

The animals' lunchtime (animals need energy too!)

<u>E</u>very <u>m</u>ouse <u>c</u>an <u>c</u>ook.
<u>G</u>ood <u>p</u>enguins <u>e</u>at <u>h</u>ot <u>m</u>ackerel and gravy.
<u>K</u>angaroos <u>have</u> <u>m</u>eat and <u>v</u>egetables <u>too</u>.

<u>E</u>very <u>m</u>ouse <u>c</u>an <u>c</u>ook.	E $m \times c \times c$	Energy / Mass × Speed of light^2
<u>G</u>ood <u>p</u>enguins <u>e</u>at <u>h</u>ot <u>m</u>ackerel and <u>g</u>ravy.	GPE $h \times m \times g$	Gravitational Potential Energy / Height × Mass × Gravity
<u>K</u>angaroos <u>have</u> <u>m</u>eat and <u>v</u>egetables <u>too</u>.	KE $½ \times m \times v^2$	Kinetic Energy / 1/2 × Mass × Velocity^2

Colin Has Energy

Colin has **legs**, a **neck** and a **h**ead.

Energy Types

L	Light Energy
E	Electrical Energy
G	Gravitational Potential Energy (GPE)
S	Sound Energy
N	Nuclear Energy
E	Elastic Potential (also called strain) Energy
C	Chemical Potential Energy
K	Kinetic Energy (KE)

H(ead) Heat Energy

Colin knows exactly where his legs, neck and head are. They are never lost! This is just like energy. It is never lost or used up. It just changes form from one type to another.

For all the different ways that we generate electrical energy there is a common theme:

With electrical energy generation all but one of the methods relies on the idea that turning an electrical generator changes kinetic energy to electrical energy.

More specifically it works like this:

Everything that burns makes heat. This means we start with chemical potential energy which we change to thermal or heat energy by burning the fuel. This thermal energy is used to heat water turning it into steam. The steam has kinetic energy and this is transferred to other objects. The steam turns a turbine (kinetic energy). The turbine is attached to an electrical generator which makes electricity. Moving water or air can also be used to turn the turbine directly.

Chemical potential energy → Thermal energy → Kinetic energy → Electrical energy

No type of energy change that we have is perfect. Some is always lost to us. It doesn't disappear, it just becomes not useful. Heat and sound are normally how energy is lost. It is important to notice that when we generate electricity we want electricity as our final product not heat and sound.

Only one method of generating electricity has no moving parts: Solar power.

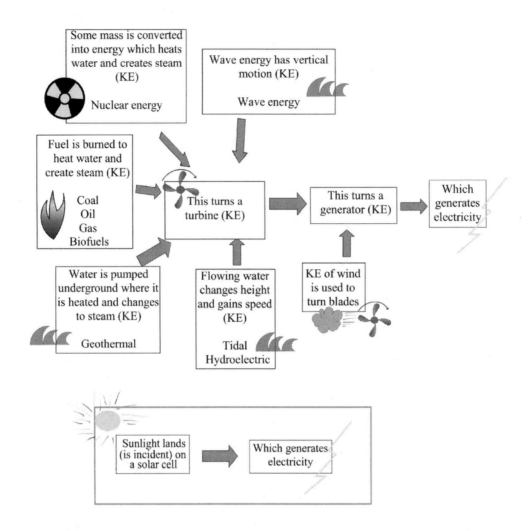

An example of lost energy is the idea of a car. If a car travels past you can hear the engine. If you touch the engine you could burn your hand but the engine has only one useful purpose and that is to make the car move. This means that the heat and sound energy represent wasted energy. Sometimes people refer to this as 'lost' energy but this is incorrect as it isn't lost. We know where it is. It just isn't very useful anymore.

Energy is Energy. We have never seen energy created or destroyed. Specifically, the total balance of mass and energy (sometimes called mass-energy) remains the same but energy can change form from one type to another.

Energy is an interesting concept in physics. It has a definition which is:

"Energy is the ability to do work"

There are a large number of different ways to calculate energy in physics depending on the scenario. We can combine them with the following rule:

Energy is never created or destroyed, it only changes forms.

And this is a rule in physics that has been tested a tremendous amount and appears, for the moment at least, unbreakable.

Examples: Kinetic Energy, Potential Energy and Energy Stored in Mass

Example 1) List the energy changes that take place in a light bulb.

Electrical energy changes to light energy. The bulb also gets hot.

$$\text{electrical energy} \rightarrow \text{light energy} + \text{heat energy}$$

Example 2) List the useful energy changes that take place in a light bulb.

The heat energy created isn't very useful (it's called a light bulb and not a heat bulb). The useful energy changes are:

$$\text{electrical energy} \rightarrow \text{light energy}.$$

Example 3) List the energy changes that take place when a large amount of rocks at the top of a mountain fall down to the base of the mountain and stop rolling.

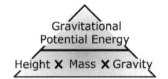 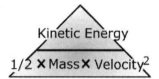

At the top of the mountain the rocks have energy because of their position (gravitational potential energy). As the rocks fall this energy because of position is changed into energy because of movement (kinetic energy).

When the rocks finally reach the ground they stop moving. Most of the energy that they had has changed into a small amount of heat and has warmed both the rocks and the ground a little. The remaining energy went into sound (they made a noise as they bounced down the side of the mountain).

Final answer: the energy changes are:
potential energy → kinetic energy → heat energy + sound energy.

Example 4) List the useful energy changes that take place in a coal fired power station.

The energy in the coal is stored as chemical potential energy. When it is burned the stored energy is released as heat. This is used to heat up water and turn it into steam. The steam has kinetic energy and this is used to turn the turbine (again kinetic energy). The turbine turns the generator and changes this kinetic energy into electric energy.

Final answer: the energy changes are:
chemical potential energy → heat energy in furnace → kinetic energy of steam → kinetic energy of turbine → electric energy from generator.

Example 5) List the energy changes that take place in a nuclear power station.

The energy in the atoms is stored as nuclear potential energy. When the nucleus is split (fission) some stored energy is released as heat. This is used to heat up water and turn it into steam. The steam has kinetic energy and this is used to turn the turbine (again kinetic energy). The turbine turns the generator and changes this kinetic energy into electric energy.

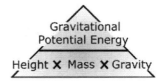

Final answer: the energy changes are:

nuclear potential energy → heat energy in reactor → kinetic energy of steam → kinetic energy of turbine → electric energy from generator.

Example 6) A 5 kg mass is at a height of 3 metres above the ground.
a) How much gravitational potential energy does it have?

GPE = height x mass x gravity
    = 3 m x 5 kg x 10 ms^{-2}
    = 150 J

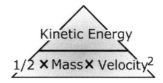

b) As it falls the gravitational potential energy is converted to kinetic energy. How much kinetic energy does the mass have at the bottom, just before it hits the ground (assume there is no air resistance)?

As it falls we assume that all of the gravitational potential energy becomes kinetic energy.

GPE → KE

The maximum value of kinetic energy is the maximum value of the gravitational potential energy!

KE = 150 J

c) What velocity does the mass have just before hitting the ground (assume air resistance is negligible).

KE = 1/2 mv^2
   = 150 J
½ mv^2 = 150 J
½ x 5 kg x v^2 = 150 J
v^2 = 60 m^{2}s^{-2}
v = √(60 m^{2}s^{-2})
  = 7.7 ms^{-1}

Questions: Kinetic Energy, Potential Energy and Energy Stored in Mass

4.1) A 20 kg object is at a height of 50 metres.
   a) How much gravitational potential energy does it have?
   b) The object is released. With what speed does it hit the ground assuming that air resistance is negligible?
   c) What would happen to this speed if air resistance was not negligible?

4.2) To run a coal fired power station I burn coal that heats water to make steam and then use this steam to turn a turbine that then creates electricity.
List the useful energy changes that are taking place.

4.3) In a nuclear power station I use the heat generated by nuclear fission to heat the water.
   a) List the useful energy changes taking place in a nuclear power station.
   b) Where does this energy come from?

4.4) List the useful energy changes taking place in each of the following.
   a) Natural gas power station
   b) Oil fired power station
   c) Hydroelectric power station (energy from falling water)
   d) Solar cell (energy from the sun)

4.5) List the energy changes that take place when a car uses its brakes to slow to a stop.

4.6) An arrow is fired from a bow.
   a) List the useful energy changes taking place.
   b) If there is 50 J of energy stored in the bow and the arrow, of mass 30 g is fired, calculate the maximum speed it will be released with.
   c) If the arrow is fired directly upwards calculate the maximum height it can achieve (ignoring air resistance).

4.7) A hydroelectric power station gets all of its energy from a lake that is 400 metres above the turbines. An average of 30 litres of water flows down the pipe every second. What is the maximum amount of electricity that could be generated every second? 1 litre of water has a mass of 1 kg.

4.8) Coal has about 30 MJ of energy per kilogram.
   a) What height would I need to lift a single litre of water to so that it had the same amount of energy?
   b) Give 2 reasons why fossil fuels are used so often.

4.9) A scientist wants to try and convert matter into pure energy.
   a) What is the amount of energy that would be released by the complete conversion of 1 kg of matter to energy?
   b) The US uses about 570 GJ per second (gigawatts). How long would the energy produced from the 1 kg be able to supply the US for?

4.10) Coal has an energy density of 30 MJ per kilogram.
   a) Assuming that an engine using coal was 100% efficient (which is impossible as energy is always lost) how much coal would be needed to lift a 30,000 kg truck a vertical height of 100 metres?
   b) Assuming the engine actually had an efficiency of 30% how much coal would now be required?

Answers: Kinetic Energy, Potential Energy and Energy Stored in Mass

4.1) A 20 kg object is at a height of 50 metres.
   a) How much gravitational potential energy does it have?
   b) The object is released. With what speed does it hit the ground assuming that air resistance is negligible?
   c) What would happen to this speed if air resistance was not negligible?

a) gravitational potential energy = height x mass x gravity
   = 50 m x 20 kg x 10 ms^{-2}
   = 10,000 J

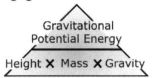

b) As the object falls the gravitational potential energy is changed to kinetic energy.

KE = GPE = 10,000 J
KE = ½ mv^2
10,000 J = ½ x 20 x v^2
v^2 = 1,000 m^{2}s^{-2}
v = √1000 m^{2}s^{-2}
   = 31.6 ms^{-1}

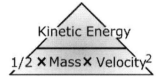

c) If the air resistance was not negligible then some of the energy would have to go to overcoming the air resistance. This would reduce the speed.

4.2) To run a coal fired power station I burn coal that heats water to make steam and then use this steam to turn a turbine that then creates electricity.
List the useful energy changes that are taking place.

Chemical potential energy → heat energy → kinetic energy → electrical energy

4.3) In a nuclear power station I use the heat generated by nuclear fission to heat the water.
   a) List the useful energy changes taking place in a nuclear power station.
   b) Where does this energy come from?

a) Nuclear energy → heat energy → kinetic energy → electrical energy

b) Answer this question by referring to E = mc^2

When the nucleus splits the 2 halves have slightly less mass. This missing mass is converted to energy using E = mc^2. The amount of energy released is given by multiplying the missing mass by the speed of light squared.

4.4) List the useful energy changes taking place in each of the following.
a) Natural gas power station
Chemical potential energy → heat energy → kinetic energy → electrical energy

b) Oil fired power station
Chemical potential energy → heat energy → kinetic energy → electrical energy

c) Hydroelectric power station (energy from falling water)
Gravitational potential energy → kinetic energy → electrical energy

d) Solar cell (energy from the sun)
Light energy → electrical energy
(solar cells have no moving parts)

4.5) List the energy changes that take place when a car uses its brakes to slow to a stop.

Kinetic energy → heat and sound

The brakes slow the wheel due to friction and this energy is converted to heating the brakes and to a smaller amount the tires and the road. This can also create a small amount of sound energy (it takes less energy to create a small sound than to heat an object).

4.6) An arrow is fired from a bow.
  a) List the useful energy changes taking place when an arrow is fired from a bow.
  b) If there is 50 J of energy stored in the bow and the arrow, of mass 30 g is fired, calculate the maximum speed it will be released with.
  c) If the arrow is fired directly upwards calculate the maximum height it can achieve (ignoring air resistance).

a) Elastic potential energy → kinetic energy

b) Elastic potential = 50 J
KE = ½ mv²
   = 50 J
Mass is measured in kilograms.
30 g = 0.03 kg
½ x 0.03 x v² = 50 J
v² = 3,333
v = 57.8 ms⁻¹

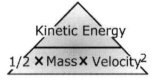

c) If the arrow is fired directly upwards then KE changes to GPE

GPE = height x mass x gravity = 50 J

height = $\dfrac{\text{GPE}}{\text{mass x gravity}}$

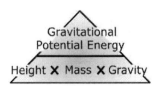

= $\dfrac{50 \text{ J}}{0.03 \text{ kg} \times 10 \text{ ms}^{-2}}$

= 167 m

4.7) A hydroelectric power station gets all of its energy from a lake that is 400 metres above the turbines. An average of 30 litres of water flows down the pipe every second. What is the maximum amount of electricity that could be generated every second? 1 litre of water has a mass of 1 kg.

First we calculate the amount of energy available per litre and then the total amount of energy available each second for the 30 litres.

GPE per litre = height x mass x gravity
= 1 kg x 10 ms^{-2} x 400 m
= 4,000 J of energy are available per litre

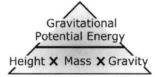

We have 30 litres travelling per second so each second we have:
= 4,000 J / litre x 30 litres / second
= 120,000 J per second.
= 120,000 watts or 120 kW

This is the maximum amount of energy that is available to be converted into electricity. In practice it will always be less than this. Nothing is 100% efficient. Energy will be lost as less useful forms such as heat and sound.

4.8) Coal has about 30 MJ of energy per kilogram.
    a) What height would I need to lift a single litre of water to so that it had the same amount of energy?
    b) Give 2 reasons why fossil fuels are used so often.

GPE = height x mass x gravity
= 30 MJ
= 30,000,000 J

height = $\dfrac{\text{GPE}}{\text{mass x gravity}}$

= $\dfrac{30,000,000 \text{ J}}{1 \text{ kg x } 10 \text{ ms}^{-2}}$

= 3,000,000 metres
= 3,000 kilometres

b) i) Fossil fuels have a large amount of chemical potential energy per kilogram.
   ii) Fossil fuels can be used directly in engines.

4.9) A scientist wants to try to convert matter into pure energy.
    a) What is the amount of energy that would be released by the complete conversion of 1 kg of matter to energy?
    b) The US uses, on average, about 570 GJ per second (gigawatts). How long would the energy produced from the 1 kg be able to supply the US for?

energy = mass x speed of light^2
= 1 kg x (3 x 10^8 ms^{-1})^2
= 9 x 10^{16} J per kilogram

b) 570 GJ = 570, 000, 000, 000 J

$$\text{time} = \frac{\text{energy produced by 1 kg}}{\text{energy used per second}}$$

$$= \frac{9 \times 10^{16} \text{ J}}{570 \times 10^9 \text{ J}}$$

= 157, 894 seconds
or 43.9 hours or 1.83 days

4.10) Coal has an energy density of 30 MJ per kilogram.
    a) Assuming that an engine using coal was 100% efficient (which is impossible as energy is always lost) how much coal would be needed to lift a 30,000 kg truck a vertical height of 100 metres?

Energy required is equal to the change in gravitational potential energy.

GPE = height x mass x gravity
      = 100 m x 30,000 kg x 10 ms^{-2}
      = 300, 000, 000 J
      = 30 MJ

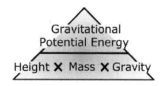

$$\frac{\text{number of}}{\text{kilograms}} = \frac{\text{energy required}}{\text{energy provided per kg}}$$

$$= \frac{30 \text{ MJ}}{30 \text{ MJ per kg}}$$

amount of coal required = 1 kg

    b) Assuming the engine actually had an efficiency of 30% how much coal would now be required?

$$\text{efficiency} = \frac{\text{useful energy out}}{\text{total energy in}}$$

$$\text{total energy in} = \frac{\text{useful energy out}}{\text{efficiency}}$$

$$= \frac{30 \text{ MJ}}{0.3}$$

$$= 100 \text{ MJ}$$

$$\frac{\text{number of}}{\text{kilograms}} = \frac{\text{energy required}}{\text{energy provided per kilogram}}$$

$$= \frac{100 \text{ MJ}}{30 \text{ MJ per kg}}$$

= 3.33 kg

Bonus Questions 4.1: Kinetic Energy, Potential Energy and Energy Stored in Mass

4.1.1) A brick has a mass of 400 g. It is inserted into a wall in a building that is being built. It is at a height of 20 m. Calculate the gravitational potential energy that the brick has at that height.

4.1.2) A workman has a mass of 80 kg. He climbs the stairs in a building from the ground floor until he reaches the floor that he is working on. When he is at the floor that he needs he is at a height of 60 m.
   a) How much has his GPE changed by?
   b) Where did the energy come from to allow him to change his height?

4.1.3) Calculate the gravitational potential energy of the following objects.
   a) A car of mass 2,000 kg at the level of the street.
   b) A plant pot of mass 4 kg at a height of 4 metres above the street.
   c) A window of mass 20 kg at a height of 100 m above the ground.

4.1.4) A mass of 20 kg is raised to a height of 12 m above the ground.
   a) Calculate the gravitational potential energy of the mass at this height.
   b) What happens if the weight is dropped and where does this energy go?

4.1.5) A tennis ball of mass 100 g is dropped from a hot air balloon at a height of 1,000 m above the ground.
   a) Ignoring air resistance, what will the speed of the ball be just before it reaches the ground?
   b) If an identical tennis ball is dropped from a height of 4,000 m, then what speed could it theoretically reach just before hitting the ground?
   c) Explain if this actually possible?

4.1.6) An arrow of mass 40 g is fired directly upwards into the air. The arrow reaches a maximum height of 60 m.
   a) Using the conservation of energy calculate the maximum initial speed that it must have had.
   b) Another arrow, this time of mass 80 g is fired directly into the air and also reaches a maximum height of 60 m. What was the initial speed of this arrow? How does this compare to the initial speed of the first arrow?

4.1.7) A ball of paper of mass 20 g is fired from a sling at a speed of 300 kmh^{-1}.
   a) What is the kinetic energy of the ball of paper?
   b) What do you expect to happen to the value of the kinetic energy? Why?

4.1.8) Calculate the kinetic energy of the following.
   a) A bicycle and its rider with a mass of 90 kg travelling at 11 ms^{-1}.
   b) A car of mass 2,000 kg travelling at a speed of 14 ms^{-1}.
   c) A fast train of mass 5×10^6 kg travelling at a speed of 230 kmh^{-1}.

4.1.9) Which of the following has the most kinetic energy?
   A truck of mass 30,000 kg travelling at 20 ms^{-1},
   A train of mass 7×10^6 kg travelling at a speed of 3 ms^{-1},
   A fighter jet of mass 12,000 kg travelling at a speed of 1,495 kmh^{-1}?

4.1.10) A ball is dropped from a window at a height of 30 m. Use the conservation of energy to calculate the velocity of the ball just before it hits the ground. Ignore the effects of air resistance.

4.1.11) A world class sprinter with a mass of 60 kg covers 100 m in 10 s.
   a) What is their average speed?
   b) Use this figure to calculate their average kinetic energy.
   c) A long distance runner with a mass of 60 kg covers 100 m in a time of 20 s. What is their average KE?
   d) How does this value compare to that of the sprinter's KE?

4.1.12) A car has a mass of 2,000 kg.
   a) Calculate the KE of the car when it is travelling at a speed of 20 ms^{-1}.
   b) A truck of mass 5,000 kg is travelling at a speed of 10 ms^{-1}. Which vehicle has the most KE?

4.1.13) Calculate the KE of a small object of mass 20 g for each of the following speeds.
   a) 10 ms^{-1}
   b) 20 ms^{-1}
   c) 40 ms^{-1}

4.1.14) In the fission of a U-235 nucleus, the nucleus splits into 2 smaller nuclei and throws out some neutrons. During this process about 0.1% of the mass changes into energy and is given out. In a nuclear explosion we see this destructive energy however in the case of a nuclear reactor we can see the positive effects that it can have generating electricity. The value of c, the speed of light, is 3 x 10^8 ms^{-1}.
   a) How many grams of mass change to energy in the fission of 1 kg of U-235?
   b) How much energy does this release?
   c) A battleship has a mass of 100 x 10^7 newtons. What height could it be lifted to using only the energy stored in 2 g of matter?

4.1.15) The world uses about 5 x 10^{20} joules of energy each year.
   a) Calculate how much mass would need to be converted to energy in a year to supply this.
   b) If the conversion of uranium 235 only converts 0.1% of its mass to this value, what mass of U-235 would be needed to produce this energy?

4.1.16) List two benefits and two drawbacks of solar power.

4.1.17) List the energy conversions that take place from when an arrow is fired directly upwards into the sky until it reaches the ground.

4.1.18) A group of scouts are sitting next to a campfire in the woods.
   a) List the energy conversions that are taking place in the fire.
   b) What is the main method of heat transfer by which they are kept warm?

4.1.19) Describe the energy changes that take place in a light bulb. State which ones are useful and which ones are not.

4.1.20) What method of energy transfer allows the sun to heat the Earth? Where else is this type of energy transfer useful?

4.1.21) A train is travelling at 70 km/hr. The train driver sees something on the track ahead and applied the brakes. The train stops. Describe the energy transfers that have occurred.

Bonus Questions 4.2: Kinetic Energy, Potential Energy and Energy Stored in Mass

4.2.1) A large bird lands on the top of a building. The bird has a mass of 500 g. The building has a height of 62 m. Calculate the gravitational potential energy of the perched bird.

4.2.2) Every year more than 9,000 people climb the CN tower in Toronto, Canada. It has a height of 446.5 metres.
    a) Assuming that the average person has a mass of 60 kg, by how much has their GPE changed?
    b) Where did the energy come from to allow them to change their height?
    c) How much energy would be required to lift all 9,000 people the 446.5 m to the observation deck of the tower.

4.2.3) Calculate the gravitational potential energy of the following objects.
    a) A truck of mass 2,500 kg that is being fixed on a ramp 2 m above the ground.
    b) An apple of mass 0.1 kg at a height of 3 metres above the street.
    c) A metal I-beam of mass 200 kg at a height of 90 m above the ground.

4.2.4) A 15 kg mass is raised to a position 14 m above the ground.
    a) Calculate the gravitational potential energy of the mass at this height.
    b) What happens if the weight is dropped and where does this energy go?

4.2.5) A small piece of ice of mass 10 g falls from the wing of an aircraft at a height of 11,000 m above the ground.
    a) Ignoring air resistance, what will the speed of the ice be just before it reaches the ground?
    b) Explain if this actually possible?

4.2.6) A ball of mass 50 g is thrown directly upwards into the air. The ball reaches a maximum height of 20 m.
    a) Using the conservation of energy, calculate the maximum initial speed that it must have had.
    b) Another ball, of mass 200 g is dropped from a height of 20 m. What speed does it attain just before hitting the ground (ignore air resistance)?

4.2.7) A plastic bag of mass 20 g is blown out of a car window. The car was travelling with a speed of 30 kmh^{-1}.
    a) What is the kinetic energy of the bag?
    b) What do you expect to happen to the value of the kinetic energy? Why?

4.2.8) Calculate the kinetic energy of the following:
    a) A jogger of mass 70 kg travelling at 3 ms^{-1}.
    b) A truck of mass 3,000 kg travelling at a speed of 14 ms^{-1}.
    c) A 1 g projectile fired from a new weapon that travels at 10 kms^{-1}.

4.2.9) Which of the following has the most kinetic energy?
        A ball of mass 0.1 kg travelling at 20 ms^{-1},
        A can of mass 1 kg travelling at 2 ms^{-1},
        A weight of mass 10 kg travelling at 0.2 ms^{-1}?

4.2.10) A plant pot falls from a window at a height of 10 m. Use the conservation of energy to calculate the velocity of the plant pot as it reaches the ground. Ignore the effects of air resistance.

4.2.11) In 1952 Frank Gianino Jr ran across America. It took him 46 days 8 hours and 36 minutes and he was about 60 kg. He covered a distance of 4,989 km during this time.
   a) What was Frank's average speed?
   b) Use this figure to calculate his average kinetic energy.
   c) How does this value compare to the average kinetic energy of a 60 kg dog walker who walks 4 km in an hour?

4.2.12) One of the things that makes space dangerous is the possibility of colliding with micrometeorites when you are in a spaceship. Calculate the KE of a meteorite of mass 10 g travelling at a speed of 60 kms^{-1}.

4.2.13) Calculate the KE of a potato of mass 200 g launched from a potato cannon at the following speeds.
   a) 30 ms^{-1}
   b) 50 ms^{-1}
   c) 80 ms^{-1}

4.2.14) The head of a pin has a mass of about 30 micrograms. The value of c, the speed of light, is 3×10^8 ms^{-1}.
   a) How much energy would be released if the head of a pin was completely converted to energy?
   b) A jumbo jet uses 140×10^6 Joules of energy every second. How long could it fly for with all of the energy from the head of the pin?

4.2.15) The sun releases about 4×10^{26} joules of energy each second. How much mass needs to be converted to energy to achieve this output?

4.2.16) List two benefits and two drawbacks of wind power.

4.2.17) What energy conversions take place when a piece of plasticine is dropped from the top of a building?

4.2.18) Coal is a source of energy that we use to create electricity.
   a) State the processes that need to occur for this to happen.
   b) What are the useful energy conversions during this process?

4.2.19) Describe the energy changes that take place in a television. State which ones are useful and which ones are not.

4.2.20) List the useful energy conversions that take place when you climb the stairs.

4.2.21) Describe the energy transfers in the car that occur when the driver applies the brakes to stop.

5. Pressure in Solids, Liquids and Gas

Dancing and Pivoting

<u>F</u>ind <u>a</u> <u>p</u>artner.
<u>P</u>artners <u>h</u>ave a <u>g</u>reat <u>d</u>ance
<u>pi</u><u>v</u>oting <u>in</u>, <u>pi</u><u>v</u>oting <u>out</u>.

<u>F</u>ind <u>a</u> <u>p</u>artner.	F $A \times P$	Force / Area × Pressure
<u>P</u>artners <u>h</u>ave a <u>g</u>reat <u>d</u>ance	P $h \times g \times \rho$	Pressure / Height × Gravity × Density
<u>pi</u><u>v</u>oting <u>in</u> <u>pi</u><u>v</u>oting <u>out</u>.	$P_1 \times V_1$ $P_2 \times V_2$	$\dfrac{Pressure_1 \times Volume_1}{Pressure_2 \times Volume_2}$

Examples: Pressure in Solids, Liquids and Gas

Example 1) A force of 10 N is exerted on an area of 2 m². What is the pressure because of the force?

pressure = $\dfrac{\text{force}}{\text{area}}$

= $\dfrac{10 \text{ N}}{2 \text{ m}^2}$

= 5 Pa

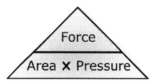

Example 2) A force of 200 newtons creates a pressure of 15 Pa. Over what area does it act?

area = $\dfrac{\text{force}}{\text{pressure}}$

= $\dfrac{200 \text{ N}}{15 \text{ Pa}}$

= 13.3 m²

Example 3) I have an area of 20 m² and I must ensure that the pressure over the area is no larger than 100 Pa. What is the maximum force that I can apply?

force = area x pressure
= 20 m² x 100 Pa
= 2,000 N

Example 4) Atmospheric pressure is 100,000 Pa.
a) What is the force due to the atmosphere on a window of area 2 m²?

force = area x pressure
= 2 m² x 100,000 Pa
= 200,000 N

b) Why doesn't the window break?

The pressure on each side of the window is the same. This means that the force on each side of the window glass is the same. There is no resultant force so the window doesn't break.

Example 5) A submarine is at a depth of 30 metres below the surface of water.
a) What is the pressure at that depth?
(Water has a density of 1,000 kg per cubic metre or 1 gram per cubic centimetre. This is something you need to know).

pressure = height x gravity x density
= 30 m x 10 ms^{-2} x 1,000 kgm^{-3}
= 300,000 Pa

b) There is an access hatch on the submarine that has an area of 0.75 m^2. What is the force on the access hatch at this depth?

force = area x pressure
= 0.75 m^2 x 300,000 Pa
= 225,000 N

Example 6) A vertical glass tube holds a liquid with a density of 500 kg per cubic metre. The tube can hold a maximum pressure difference of 40,000 Pa. What is the maximum height of the container?

height = $\dfrac{\text{pressure}}{\text{gravity x density}}$

= $\dfrac{40,000 \text{ Pa}}{10 \text{ ms}^{-2} \text{ x } 500 \text{ kgm}^{-3}}$

= 8 m

Example 7) On a new planet some scientists measure the pressure 10 metres under water (density 1,000 kgm^{-3}) to be 50,000 Pa. What is the value of gravity on that planet?

gravity = $\dfrac{\text{pressure}}{\text{height x density}}$

= $\dfrac{50,000 \text{ Pa}}{10 \text{ m x } 1,000 \text{ kgm}^{-3}}$

= 5 ms^{-2}

Example 8) I have a statue standing on a plinth (a large base). The statue has a weight of 30,000 newtons and a base of 10 m^2. What is the pressure under the base due to the statue?

pressure = $\dfrac{\text{force}}{\text{area}}$

= $\dfrac{30,000 \text{ N}}{10 \text{ m}^2}$

= 3,000 Pa

Example 9) A tree of mass of 2,000 kg is placed on a square steel base with sides of 6 m. What pressure does the tree generate beneath the steel plate?

weight = mass x gravity
= 2,000 kg x 10 ms^{-2}
= 20,000 N

The force on the steel base is the weight of the tree.

pressure = $\dfrac{\text{force}}{\text{area}}$

The area of the steel plate is given by

area = 6 m x 6 m
= 36 m^2

pressure = $\dfrac{20,000 \text{ N}}{36 \text{ m}^2}$
= 556 Pa

Example 10) I have some gas in a cylinder with a volume of 2 m^3 and a pressure of 10,000 Pa. I compress the gas in the cylinder until it has a volume of 0.1 m^3. What is the new pressure?

$P_2 = \dfrac{P_1 V_1}{V_2}$

$= \dfrac{10,000 \text{ Pa} \times 2 \text{ m}^3}{0.1 \text{ m}^3}$

= 200,000 Pa
= 200 kPa

$\boxed{\dfrac{\text{Pressure}_1 \times \text{Volume}_1}{\text{Pressure}_2 \times \text{Volume}_2}}$

Example 11) I have a mass of gas with a volume of 12 m^3 and pressure 3,000 Pa. I wish to compress the gas but the container that it is in cannot safely hold a pressure greater than 300,000 Pa. What is the smallest volume the gas can safely occupy?

$V_2 = \dfrac{P_1 V_1}{P_2}$

$= \dfrac{3,000 \text{ Pa} \times 12 \text{ m}^3}{300,000 \text{ Pa}}$

= 0.12 m^3

$\boxed{\dfrac{\text{Pressure}_1 \times \text{Volume}_1}{\text{Pressure}_2 \times \text{Volume}_2}}$

Questions: Pressure in Solids, Liquids and Gas

5.1) Some plants of mass 3 kg are placed on a stand of area 0.0025 m².
   a) What is the pressure created by the weight of the plants under the stand?
   b) Over time the stand slowly sinks into the ground. What can be done to prevent this from happening?

5.2) The pressure due to Earth's atmosphere is 1 x 10⁵ Pa. What depth would you have to go to underwater to equal this pressure due to water? What would the total pressure be at this depth?

5.3) In a hydraulic lift (This is a type force multiplier. These are discussed further at the end of the book, inside the complicated systems section) 25 N of force are supplied to an area of 0.05 m². This liquid is connected to another surface which has an area of 1 m². What is the force created by this pressure on the second surface?

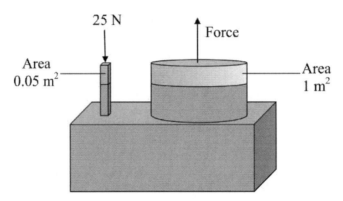

5.4) An elephant of mass 1,000 kg has 4 feet. Each foot has an area of 0.1 m².
   a) What is the pressure under the elephant's feet?
   b) A person of mass 50 kg is wearing a pair of high heeled shoes. They lean back slightly so that all of their weight is supported on the heels of the shoes. The heels each have an area of 1 cm². Calculate the pressure that is created by the person's weight.

5.5) Explain why tractor tires are so large.

5.6) Atmospheric pressure is 100,000 Pa. What is the force due to atmospheric pressure on a wall that is 10 metres long and 3 metres high? Why doesn't the wall fall over?

5.7) There is a pipe running from a height of 300 metres down to a hydroelectric power station. The power station needs to close the valve at the bottom of the pipe to do some maintenance work. The pipe has a cross sectional area of 0.5 m² and it is full of water. What is the force on the valve at the power station?

5.8) A high tech submarine can safely handle pressures of up to 40 MPa. What is the maximum depth that it can dive to underwater?

5.9) A gas is contained in a piston. The gas is at a pressure of 10 kPa and the volume is 2 m³. The volume is slowly reduced to 0.2 m³. What is the new pressure?

5.10) A gas is at a pressure of 30 MPa and is contained in a volume of 3 m³. The volume is increased until the pressure of the gas is 6 MPa. What is the new volume of the gas?

5.11) A gas has an initial volume of 2.5 m³. This is increased until its volume is 14 m³ and its pressure is 1 kPa. What was its original pressure?

Answers: Pressure in Solids, Liquids and Gas

5.1) Some plants of mass 3 kg are placed on a stand of area 0.25 m².
   a) What is the pressure created by the weight of the plants under the stand?
   b) Over time the stand slowly sinks into the ground. What can be done to prevent this from happening?

a) $\text{pressure} = \dfrac{\text{force}}{\text{area}}$

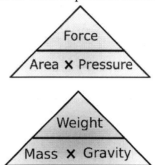

The force is from the weight of the plants.

weight = mass x gravity
= 3 kg x 10 ms⁻²
= 30 N

$\text{pressure} = \dfrac{30 \text{ N}}{0.25 \text{ m}^2}$

= 120 Pa

b) The plants sink into the ground as the pressure at the base is too big. The pressure on the ground will become smaller as the area at the base increases. Therefore, to stop the stand sinking, increase the area of the base. Putting the stand on something (like wooden blocks) with an area greater than 0.25 m² will reduce the pressure at the base. More area = less pressure.

5.2) The pressure due to Earth's atmosphere is 1 x 10⁵ Pa. What depth would you have to go to underwater to equal this pressure due to water? What would the total pressure be at this depth?

$\text{height} = \dfrac{\text{pressure}}{\text{gravity x density}}$

$= \dfrac{1 \times 10^5 \text{ Pa}}{10 \text{ ms}^{-2} \times 1{,}000 \text{ kgm}^{-3}}$

= 10 m

You would need to be at a depth of 10 m to experience a pressure due to the water that is the same magnitude as normal atmospheric pressure.

If you went to a depth of 10 m underwater, you would experience a pressure of 1 x 10⁵ Pa due to the water and another 1 x 10⁵ Pa due to the atmosphere above you. The total pressure would be 2 x 10⁵ Pa.

5.3) In a hydraulic lift (This is a type force multiplier. These are discussed further at the end of the book, inside the complicated systems section.) 25 N of force are supplied to an area of 0.05 m². This liquid is connected to another surface which has an area of 1 m². What is the force created by this pressure on the second surface?

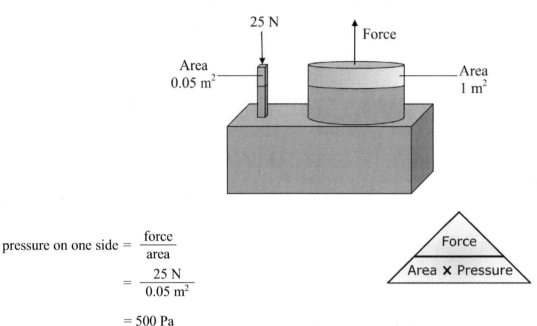

pressure on one side = $\dfrac{\text{force}}{\text{area}}$

= $\dfrac{25 \text{ N}}{0.05 \text{ m}^2}$

= 500 Pa

pressure on the other side is also 500 Pa.
force = pressure x area
= 500 Pa x 1 m²
= 500 N

5.4) An elephant of mass 1,000 kg has 4 feet. Each foot has an area of 0.1 m².

   a) What is the pressure under the elephant's feet?
   b) A person of mass 50 kg is wearing a pair of high heeled shoes. They lean back slightly so that all of their weight is supported on the heels of the shoes. The heels each have an area of 1 cm². Calculate the pressure that is created by the person's weight.

a) The total area of the elephant's feet is 4 x 0.1 m² = 0.4 m²
The force from the mass of the elephant is its weight.

weight = mass x gravity
= 1,000 kg x 10 ms⁻²
= 10,000 N

pressure = $\dfrac{\text{force}}{\text{area}}$

= $\dfrac{10,000 \text{ N}}{0.4 \text{ m}^2}$

= 2,500 Pa

b) weight = mass x gravity
= 50 kg x 10 ms^{-2}
= 500 N

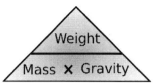

Area is measured in m^2. There are 10,000 cm^2 in 1 m^2. I can find out the number of cm^2 in 1 m^2 with the following calculation.

A square which is 1m x 1m is also 100 cm x 100 cm.

This means that
1 m x 1 m = 100 cm x 100 cm

Which becomes
1 m^2 = 10,000 cm^2
The area of interest is 2 cm^2

So to convert it from cm^2 to m^2 we perform the following calculation.

area in m^2 = $\dfrac{2 \text{ cm}^2}{10,000 \text{ cm}^2}$

= 0.0002 m^2

pressure = $\dfrac{\text{force}}{\text{area}}$

= $\dfrac{500 \text{ N}}{0.0002 \text{ m}^2}$

= 2,500,000 Pa

This means that the pressure under each high heel is 1,000 times greater than the pressure under each of the elephant's feet!

5.5) Explain why tractor tires are so large.

Something happens:
The tires on a tractor are much larger than normal tires.

Which has an effect:
This means that there is a larger area than normal in contact with the ground.

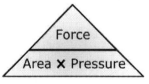

Which means:
As pressure = force / area, if the area is larger then the pressure will be smaller. There is less pressure on the ground from the weight of the tractor because of the size of the wheels.

Therefore:
The tractor is able to travel over much softer ground without sinking.

Final Answer:
The tires on a tractor are much larger than normal tires. This means that there is a larger area than normal in contact with the ground. As pressure = force / area, if the area is larger then the pressure will be smaller. There is less pressure from the weight of the tractor because of the size of the wheels. The tractor is able to travel over much softer ground without sinking.

5.6) Atmospheric pressure is 100,000 Pa. What is the force due to atmospheric pressure on a wall that is 10 metres long and 3 metres high? Why doesn't the wall fall over?

force = area x pressure
area = length x height
    = 10 m x 3 m
    = 30 m²
force = 30 m² x 100,000 Pa
    = 3,000,000 N

The wall does not fall over because there is the same amount of force due to the atmosphere at the other side (the wall is surrounded by the atmosphere).

5.7) There is a pipe running from a height of 300 metres down to a hydroelectric power station. The power station needs to close the valve at the bottom of the pipe to do some maintenance work. The pipe has a cross sectional area of 0.5 m² and it is full of water. What is the force on the valve at the power station?

pressure = height x gravity x density
    = 300 m x 10 ms⁻² x 1,000 kgm⁻³
    = 3,000,000 Pa
    = 3 MPa

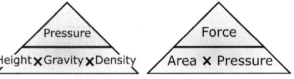

force = area x pressure
    = 0.5 m² x 3,000,000 Pa
    = 1,500,000 N
    = 1.5 MN

5.8) A high tech submarine can safely handle pressures of up to 40 MPa. What is the maximum depth that it can dive to underwater?

We want to calculate the height. We know the maximum pressure and also the value for density and gravity. We will need the equation to calculate pressure in a liquid.

40 MPa = 40,000,000 Pa

height = $\dfrac{\text{pressure}}{\text{gravity x density}}$

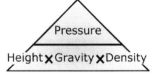

= $\dfrac{40,000,000 \text{ Pa}}{10 \text{ ms}^{-2} \text{ x } 1,000 \text{ kgm}^{-3}}$

= 4,000 m
= 4 km

The submarine can reach a maximum depth of 4 kilometres.

5.9) A gas is contained in a piston. The gas is at a pressure of 10 kPa and the volume is 2 m³. The volume is slowly reduced to 0.2 m³. What is the new pressure?

$$P_2 = \frac{P_1 V_1}{V_2}$$

$$= \frac{10 \text{ kPa} \times 2 \text{ m}^3}{0.2 \text{ m}^3}$$

$$= 100 \text{ kPa}$$

$$\frac{\text{Pressure}_1 \times \text{Volume}_1}{\text{Pressure}_2 \times \text{Volume}_2}$$

5.10) A gas is at a pressure of 30 MPa and is contained in a volume of 3 m³. The volume is increased until the pressure of the gas is 6 MPa. What is the new volume of the gas?

$$V_2 = \frac{P_1 V_1}{P_2}$$

$$= \frac{30 \text{ MPa} \times 3 \text{ m}^3}{6 \text{ MPa}}$$

$$= 15 \text{ m}^3$$

$$\frac{\text{Pressure}_1 \times \text{Volume}_1}{\text{Pressure}_2 \times \text{Volume}_2}$$

5.11) A gas has an initial volume of 2.5 m³. This is increased until its volume is 14 m³ and its pressure is 1 kPa. What was its original pressure?

$$P_1 = \frac{P_2 V_2}{V_1}$$

$$= \frac{1 \text{ kPa} \times 14 \text{ m}^3}{2.5 \text{ m}^3}$$

$$= 5.6 \text{ kPa}$$

$$\frac{\text{Pressure}_1 \times \text{Volume}_1}{\text{Pressure}_2 \times \text{Volume}_2}$$

Bonus Questions 5.1: Pressure in Solids, Liquids and Gas

5.1.1) Calculate the pressure beneath a square metal plate with sides 20 cm long and a weight of 500 N.

5.1.2) What is the pressure beneath a round base of mass 3 kg that is supporting a plant of mass 12 kg? The base has a radius of 30 cm.

5.1.3) A camel has a foot that can be approximated by a circle of radius 15 cm. Calculate the pressure from the camel if it has a mass of 600 kg and all of its feet are on the ground.

5.1.4) The largest land mammal in the world is the elephant. An elephant has a mass of up to 6,000 kg. Calculate the pressure that the elephant generates if each of its feet has a surface area of 1,600 cm^2.

5.1.5) A lady is wearing shoes that have a square heel of width 1 cm. Calculate the pressure that exists if she leans back slightly so that only the heels of both of her shoes support her weight. Assume that she has a mass of 50 kg.

5.1.6) An object of mass 3,000 kg has a pressure beneath it of 300 Pa. What area is in contact with the ground to support the weight of the object?

5.1.7) An object has an area of 0.6 m^2 with a pressure beneath it of 1,200 Pa.
   a) Find the total force acting on the object.
   b) If this force comes entirely from the weight of the object calculate the mass that the object has.

5.1.8) Calculate the pressure from a person of mass 60 kg standing on the ground with both feet. Each foot has an area of 400 cm^2

5.1.9) If the person in the previous question is standing on an inflatable air mattress without their feet touching the ground (the mattress is fully supporting their weight) then the mattress must be matching the pressure that the person is producing. If the air mattress has a width of 1.5 m and a length of 2.0 m, then what is the maximum force that the air mattress could possibly produce with this internal pressure?

5.1.10) Could the air mattress in the previous question theoretically produce enough force to lift the following?
   a) A car of mass 2,000 kg
   b) A truck of mass 4,000 kg
   c) A bus of mass 8,000 kg
   d) Comment on how the air mattress would have to be located to perform these feats. (It is not a good idea to try this as it may damage the vehicle. This is because the parts under the car are not designed for a large upward force. Also if the vehicle exhaust or brakes or any other part is hot the mattress could catch fire!)

5.1.11) The pressure is measured under the surface of a liquid. If the pressure is 400,000 Pa and the pressure due to the atmosphere is 100,000 Pa at what depth is the pressure being measured?
The density of the liquid is 1,200 kgm^{-3}.

5.1.12) What is the density of a liquid if, at a depth of 30 m the pressure is 240,000 Pa more than it is at the surface?

5.1.13) Calculate the pressure due to a liquid of density 1,320 kgm^{-3} at a depth of 12 m below the surface of the liquid.

5.1.14) Consider the following liquids. Calculate the pressures at each of the following points and identify at which point the pressure is the greatest and where it is the least.
   a) Depth = 25 m, density = 500 kgm^{-3}
   b) Depth = 12 m, density = 1,000 kgm^{-3}
   c) Depth = 780 mm, density = 13,560 kgm^{-3}
   d) Depth = 800 cm, density = 8,000 kgm^{-3}

5.1.15) Calculate the pressure at a depth of 3 m in liquids of the following densities.
   a) 3.4 gcm^{-3}
   b) 0.2 gcm^{-3}
   c) 1 gcm^{-3}

5.1.16) A cube of side 1 m is lowered into a liquid of density 820 kgm^{-3}.
   a) What is the total area of the cube?
   b) What is the pressure at the top face of the cube when the centre of the cube is at a depth of 3.5 m?
   c) What is the force pushing down on the cube's top face because of this pressure?
   d) What is the pressure at the bottom face when the centre of the cube is at a depth of 3.5 m?
   e) What is the force pushing upwards on the cube's base because of this pressure?
   f) What is the difference between the force pushing downwards at the top and the force pushing upwards on the base of the cube?

5.1.17) What is the pressure due to the water on a flat piece of plastic that has sunk into the water to a depth of 100 m? If the plastic has an area of 3 m^2 what force is acting on it due to the pressure of the water? Why does the plastic not get completely destroyed?

5.1.18) A huge balloon filled with gas has a volume of 300 m^3 when it is at an external pressure of 200 kPa.
   a) By what proportion will the volume change if the external pressure is doubled?
   b) What about if the external pressure is increased by a factor of 10?

5.1.19) At a height of 1 m above the ground and at a pressure of 100,000 Pa the balloon has a volume of 0.3 m^3.
   a) What volume would the balloon have when it is released and rises 10 km into the air where the air pressure is approximately 26,500 Pa? (This is the approximate height of Mount Everest.)
   b) What could happen to the balloon as the pressure decreases?

5.1.20) A military submarine can travel to a depth of 600 m beneath the surface of the ocean. The volume of the air inside the submarine is 15,000 m^3 and it is at normal atmospheric pressure. What volume would the air have if, instead of being in a submarine, it was in a giant piston at the same depth? The density of water is 1,000 kgm^{-3}. Normal atmospheric pressure is 100,000 Pa and the pressure under the surface of the water is equal to atmospheric pressure + the pressure from the water.

Bonus Questions 5.2: Pressure in Solids, Liquids and Gas

5.2.1) A concrete plinth has a weight of 500 N. If it has sides of length 40 cm and 60 cm, calculate the pressure beneath it.

5.2.2) A statue of mass 300 kg sits atop a round base. The base has a radius of 30 cm. What is the pressure beneath the base due to the statue?

5.2.3) A young and growing tree is placed into a large plant pot. Initially the tree and pot have a mass of 12 kg.
   a) If the base of the plant pot is 20 cm by 30 cm, calculate the pressure on the ground due to the tree and pot.
   b) Over time what happens to the pressure under the plant pot?
   c) Where does the extra mass, if any, come from?

5.2.4) A polar bear is sitting on its bottom on the snow. Polar bears are the largest land carnivores in the world and this one has a mass of 650 kg. Assuming that the polar bear's behind can be approximated by a circle of diameter 1.2 m, calculate the pressure on the snow from the polar bear.

5.2.5) An object of mass 1,200 kg has a pressure beneath it of 420 Pa. What area is in contact with the ground to support the weight of the object?

5.2.6) A large dog has a mass of 20 kg. The pressure under the dog is 6,370 Pa. Assume that the paws can be modelled as identical circles. What is the radius of each paw?

5.2.7) A cabinet has an area of 0.8 m^2 with a pressure beneath it of 1,500 Pa.
   a) Find the total force acting on the ground.
   b) If this force comes entirely from the weight of the cabinet, calculate the mass that the cabinet has.
   c) What is the force from the ground acting on the cabinet?

5.2.8) A person of mass 70 kg is sitting on a three legged bar stool. If the bar stool has a mass of 5 kg, calculate the pressure on the ground under each foot. Each foot of the stool has an area of 20 cm^2.

5.2.9) A boat of mass 1,200 kg has been beached on a sunny day so that the owners can go for a walk. The owners didn't realise however that half of the weight of the boat is being supported on a small flat rock which has an area of 24 cm^2 in contact with the boat. Calculate the pressure on the boat from the rock.

5.2.10) A house is being jacked up onto a support beam while some work can be performed on the foundations. The maximum pressure that the house can endure in any one spot before serious structural damage will occur is 1×10^7 Pa. Each support has an area of 100 cm^2. Calculate the number of supports required for a house of mass 60,000 kg.

5.2.11) A diver decides to measure the total pressure at a point underwater. If the pressure measured is 250,000 Pa and the pressure due to the atmosphere is 100,000 Pa, at what depth is the pressure being measured? The density of the water is 1,000 kgm^{-3}.

5.2.12) 30 cm under the surface of a liquid the pressure is 24,000 Pa greater than it is at the surface. What is the density of the liquid?

5.2.13) What is the pressure due to a liquid of density 1,720 kgm^{-3} at a depth of 0.8 m below the surface?

5.2.14) At which position in the following liquids is the pressure the greatest?
   a) Depth = 12.5 m, density = 300 kgm^{-3}
   b) Depth = 8 m, density = 500 kgm^{-3}
   c) Depth = 78 cm, density = 13,560 kgm^{-3}
   d) Depth = 91 cm, density = 8,400 kgm^{-3}

5.2.15) Calculate the pressure at a depth of 12 metres in liquids of the following densities.
   a) 3.2 gcm^{-3}
   b) 0.6 gcm^{-3}
   c) 1 gcm^{-3}

5.2.16) A cube with sides that are each 1.5 metres long is lowered into a liquid of density 1,220 kgm^{-3}.
   a) What is the total area of the cube?
   b) What is the pressure at the top face of the cube when the centre of the cube is at a depth of 2.25 m?
   c) What is the force pushing down on the cube's top face because of this pressure?
   d) What is the pressure at the bottom face when the centre of the cube is at a depth of 2.25 m?
   e) What is the force pushing upwards on the cube's base because of this pressure?
   f) What is the difference between the force pushing downwards at the top and the force pushing upwards on the base of the cube?

5.2.17) A square pane of glass with sides that are each 1 m is lowered 20 m below the surface of water.
   a) What is the pressure due to the water on the piece of glass?
   b) What is the force that is acting on it due to the pressure of the water?
   c) Why does the glass not break?

5.2.18) A balloon is filled with 1 m^3 of a gas and has a pressure of 1 atm. It is able to change size freely. Calculate the volume of the balloon when it is at the following locations.
   a) The top of Ben Lomond (0.9 atm)
   b) The top of Mont Blanc (0.5 atm)
   c) 10 metres under the surface of the sea. (1 atm = 100,000 Pa)

5.2.19) A large weather balloon filled with helium is released with a camera suspended from it. The balloon rises into the air and continues to rise. After a while the balloon and camera return to the earth. What is it that happens that ensures the balloon and camera return and why?

5.2.20) A balloon holding 1 m^3 of air at atmospheric pressure is weighted down and released so that it floats down to the bottom of the Marianas Trench. Calculate the volume of the balloon when it reaches the sea floor. The depth of the trench is 10,994 m. The density of water is 1,000 kgm^{-3}. Normal atmospheric pressure is 100,000 Pa and the pressure under the surface of the water is equal to atmospheric pressure + the pressure from the water.

6. Work, Energy, Power and Heat

Colin Changes Tune

Every morning Colin changes tune.
We find dancing
will take place.

Every	E	
morning **C**olin **changes** **t**une.	m x c x ΔT (Δ = change)	Energy = Mass × Specific Heat Capacity × Change in Temperature
We find **d**ancing	W F x d	Work Done = Force × Distance
will **t**ake **p**lace.	W t x P	Work Done = Time × Power

Examples: Work, Energy, Power and Heat

Before we progress in this section it is worth making a note about the 2 main types of temperature scale that are used in physics. They are Celsius and kelvin. They are identical except the value of a temperature in kelvin is 273 degrees bigger than its value in Celsius.

These are both different to the Fahrenheit scale. We do not use the Fahrenheit scale any more for physics calculations. This is because of the changes that mean much of the world now uses kelvin and Celsius and physics requires communication between groups of researchers. For this to happen we need a common language to occur.

The common language of physics includes metres (for distance), kelvin (for temperature), seconds (for time), kilogram (for mass), ampere (for current), candela (for luminosity) and mole (for amount of a substance). The last two of these you will not meet in this part of physics although the mole is very important in chemistry.

The kelvin scale starts at absolute zero. 0 K is as cold as it can possibly get. There is a reason that temperature cannot be colder than this. Temperature is related to tiny random vibrations of the molecules in a substance. Thus temperature is related to the random kinetic energy of the molecules of a substance. The lowest temperature available is when the molecules are not vibrating. At this temperature they have no random kinetic energy. This is 0 kelvin.

The Celsius scale is defined between 0 degrees Celsius (pure melting ice) and 100 degrees Celsius (pure boiling water).

A difference in temperature of 30 degrees is the same in kelvin and Celsius.
Let's examine the difference between 20 degrees Celsius and 0 degrees Celsius.

In Celsius it is just a difference of 20 degrees - 0 degrees = 20 degrees Celsius.
In kelvin it is (20 degrees + 273) - (0 degrees + 273) = 293 – 273 = 20 kelvin.

A difference of 5 degrees Celsius is the same as a difference of 5 kelvin. The specific heat capacity equation uses kelvin, but as we have just seen, for temperature differences the scales are identical.

In terms of actual temperatures, 30°C (86°F) might be nice weather for sunbathing but I wouldn't go sunbathing at 30 K (it's -243°C, which is - 405°F)!

Another thing that it is important to be aware of is that the specific heat capacity is often written as the small letter c. Just for a little confusion this is the same letter that is used for the speed of light. This is why the equation for energy in heating is E = m x c x change in temperature. This is not the same as E = m x c^2 which is the energy stored in mass and is not related to heating!

In this text we will use SHC for specific heat capacity to avoid confusion.

Example 1) I have 1 kg of water (specific heat capacity 4.2 kJkg^{-1}K^{-1}). If I wish to heat it by 30 degrees Celsius how much energy do I need to add to it?

energy = mass x SHC x change in temperature
= 1 kg x 4,200 Jkg^{-1}K^{-1} x 30 K
= 126,000 J
= 126 kJ

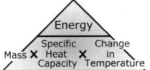

Example 2) I have 300 ml of water (specific heat capacity 4.2 kJkg^{-1}K^{-1}). If I wish to heat it by 30 degrees Celsius how much energy do I need to add to it?

energy = mass x SHC x change in temperature
= 0.3 kg x 4,200 Jkg^{-1}K^{-1} x 30 K
= 37,800 J
= 37.8 kJ

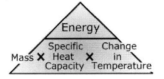

Example 3) How much energy is required if I wish to heat 1 kg of water (specific heat capacity 4.2 kJkg^{-1}K^{-1}) from 0 degrees Celsius to 100 degrees Celsius?

energy = mass x SHC x change in temperature
= 1 kg x 4,200 Jkg^{-1}K^{-1} x 100 K
= 420,000 J
= 420 kJ

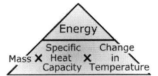

Example 4) 300 J of heat energy is added to 1 kg of water (specific heat capacity 4.2 kJkg^{-1}K^{-1}) at 30 degrees Celsius. What is its new temperature?

$$\text{change in temperature} = \frac{\text{energy}}{\text{mass x SHC}}$$

$$= \frac{300 \text{ J}}{1 \text{ kg x } 4{,}200 \text{ Jkg}^{-1}\text{K}^{-1}}$$

$$= 0.07 \text{ K}$$

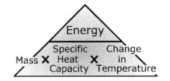

final temperature = initial temperature + change in temperature
= 30 degrees Celsius + 0.07 degrees Celsius
= 30.07 degrees Celsius
= 30.1 degrees Celsius

Example 5) I give 10,000 J of thermal energy to 1 kg of an unknown liquid. Its temperature is raised by 15 degrees. What is its specific heat capacity?

$$\text{SHC} = \frac{\text{energy}}{\text{mass x change in temperature}}$$

$$= \frac{10{,}000 \text{ J}}{1 \text{ kg x } 15 \text{ K}}$$

$$= 667 \text{ Jkg}^{-1}\text{K}^{-1}$$

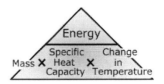

Example 6) I add 50,000 J of energy to 500 g of an unknown liquid. Its temperature is raised by 32 kelvin. What is its specific heat capacity?

SHC $= \dfrac{\text{energy}}{\text{mass x change in temperature}}$

$= \dfrac{50{,}000 \text{ J}}{0.5 \text{ kg} \times 32 \text{ K}}$

$= 3{,}125 \text{ Jkg}^{-1}\text{K}^{-1}$

$= 3.13 \text{ kJkg}^{-1}\text{K}^{-1}$

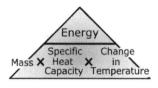

Example 7) I add 3,000 J of energy to an amount of water (specific heat capacity 4.2 kJkg^{-1}K^{-1}). Its temperature is raised by 10 degrees. How much water is there?

mass $= \dfrac{\text{energy}}{\text{SHC x change in temperature}}$

$= \dfrac{3{,}000 \text{ J}}{4{,}200 \text{ Jkg}^{-1}\text{K}^{-1} \times 10 \text{ K}}$

$= 0.071 \text{ kg}$
$= 71 \text{ g}$

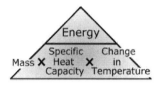

Example 8) A worker pushes a box. He uses a force of 200 N and pushes the box for a distance of 300 metres. How much work does he perform?

work done = force x distance
 = 200 N x 300 m
 = 60,000 J
 = 60 kJ

Example 9) A motorist is pushing their car uphill. They need to push it to the top of the hill. The force they push with is 500 newtons. After they have performed 12,000 J of work they stop pushing the car as they have reached the top of the hill. How far was it from when they started to push their car to the top of the hill?

distance $= \dfrac{\text{work done}}{\text{force}}$

$= \dfrac{12{,}000 \text{ J}}{500 \text{ N}}$

$= 24 \text{ m}$

Example 10) A man works for a removal company. He has to lift a 20 kg box a distance of 1.2 metres vertically.
a) How much work does he need to do?

The force here is equal to the weight of the box (the box is being lifted).

weight = mass x gravity
 = 20 kg x 10 ms^{-2}
 = 200 N

The distance here is the vertical height the box is lifted.

work done = force x distance
 = 200 N x 1.2 m
 = 240 J

b) He lifts the box in a time of 1.3 seconds. How much power does he generate?

power = $\dfrac{\text{work done}}{\text{time}}$

 = $\dfrac{240 \text{ J}}{1.3 \text{ s}}$

 = 185 J/s
 = 185 watts

Example 11) A train is generating a force of 50 kN. It travels with a speed of 30 ms^{-1}.
a) What is the distance that the train travels in 1 second?

distance = time x speed
 = 1 s x 30 ms^{-1}
 = 30 m

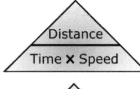

b) What is the work done in that time?

work done = force x distance
 = 50 kN x 30 m
 = 1,500 kJ
 = 1.5 MJ

c) What is the power generated in that second?

power = $\dfrac{\text{work done}}{\text{time}}$

 = $\dfrac{1,500,000 \text{ J}}{1 \text{ s}}$

 = 1,500,000 J/s
 = 1.5 MW

Questions: Work, Energy, Power and Heat

6.1) 5,000 J of energy is added to 1 kilogram of an unknown liquid. Its temperature rises by 5 degrees Celsius. Calculate the specific heat capacity of the liquid.

6.2) 3 kJ of energy is added to 100 g of a liquid. The temperature of the liquid changes by 12 kelvin. What is the specific heat capacity of the liquid?

6.3) Energy is added at a rate of 100 watts to a liquid of specific heat capacity 3 kJkg^{-1}K^{-1} and mass 300 g. How much time will be needed for the temperature of the liquid to rise by 30 degrees Celsius?

6.4) I add 50 watts of energy to a liquid of specific heat capacity 780 Jkg^{-1}K^{-1} for 5 minutes. The temperature of the liquid changes by 4 degrees Celsius. What is the mass of the liquid?

6.5) The initial temperature of a glass of water (specific heat capacity 4.2 kJkg^{-1}K^{-1}) is 35 degrees Celsius. It has a mass of 250 g. 5,000 J of thermal energy is supplied to the glass of water.
   a) Calculate the new temperature of the water.
   b) When the temperature is measured it is less than expected. Give 2 reasons for this.

6.6) I push a 10 kg box with a force of 200 N from one end of a corridor to the other end, 30 metres away. How much work have I done?

6.7) A cart requires a force of 300 N to move it. The cart is pushed for a while until the child pushing it gets tired. By the time the child has stopped to rest he has performed 3,750 J of work.
which point the pressure is the greatest and where it is the least.
   a) How far has the cart been pushed?
   b) The total time that the child spent pushing the cart was 1 minute. How much power did he generate?

6.8) An elevator moves a 2,000 kg car 10 metres vertically upwards.
   a) How much work is done?
   b) The car is lifted the 10 metres in 40 seconds. What power does the elevator generate?

6.9) A train engine is producing power at a rate of 100 kW. This is how it is able to move forward. The train is travelling at a constant speed. How much work is done over 1 minute by the train engine?

6.10) A car engine produces a force of 10 kN when it is moving at 25 ms^{-1}.
   a) What power does it generate?
   b) If 1 horsepower is 746 watts how many hp does the car generate?

Answers: Work, Energy, Power and Heat

6.1) 5,000 J of energy is added to 1 kilogram of an unknown liquid. Its temperature rises by 5 degrees Celsius. Calculate the specific heat capacity of the liquid.

$$\text{SHC} = \frac{\text{Energy}}{\text{mass} \times \text{change in temperature}}$$

$$= \frac{5,000 \text{ J}}{1 \text{ kg} \times 5 \text{ K}}$$

$$= 1,000 \text{ Jkg}^{-1}\text{K}^{-1}$$
$$= 1.0 \text{ kJkg}^{-1}\text{K}^{-1}$$

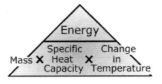

6.2) 3 kJ of energy is added to 100 g of a liquid. The temperature of the liquid changes by 12 kelvin. What is the specific heat capacity of the liquid?

$$\text{SHC} = \frac{\text{energy}}{\text{mass} \times \text{change in temperature}}$$

mass is measured in kilograms: 100 g = 0.1 kg

$$\text{SHC} = \frac{3,000 \text{ J}}{0.1 \text{ kg} \times 12 \text{ K}}$$

$$= 2,500 \text{ Jkg}^{-1}\text{K}^{-1}$$
$$= 2.5 \text{ kJkg}^{-1}\text{K}^{-1}$$

6.3) Energy is added at a rate of 100 watts to a liquid of specific heat capacity 3 kJkg^{-1}K^{-1} and mass 300 g. How much time will be needed for the temperature of the liquid to rise by 30 degrees Celsius?

First we must calculate the energy required to heat the liquid. Then we can calculate the length of time we will need to have to supply this amount of energy at 100 J per second.

energy = mass x specific heat capacity x change in temperature

mass is measured in kilograms.
300 grams = 0.3 kg
energy = 0.3 kg x 3 kJkg^{-1}K^{-1} x 30 K
= 0.3 kg x 3,000 Jkg^{-1}K^{-1} x 30 K
= 27,000 J

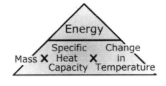

To heat the liquid, we need to do 27,000 J of work on it.

$$\text{time} = \frac{\text{work done}}{\text{power}}$$

$$= \frac{27,000 \text{ J}}{100 \text{ W}}$$
$$= 270 \text{ seconds}$$

6.4) I add 50 watts of energy to a liquid of specific heat capacity 780 Jkg^{-1}K^{-1} for 5 minutes. The temperature of the liquid changes by 4 degrees Celsius. What is the mass of the liquid?

work done in heating = time x power
time is measured in seconds.
work done = 5 min x 60 s/min x 50 J/s
 = 15,000 J
 = 15 kJ

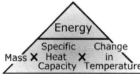

$$\text{mass} = \frac{\text{energy}}{\text{SHC x change in temperature}}$$

$$= \frac{15{,}000 \text{ J}}{780 \text{ J kg}^{-1}\text{K}^{-1} \times 4 \text{ K}}$$

= 4.81 kg

6.5) The initial temperature of a glass of water (specific heat capacity 4.2 kJkg^{-1}K^{-1}) is 35 degrees Celsius. It has a mass of 250 g. 5,000 J of thermal energy is supplied to the glass of water.
  a) Calculate the new temperature of the water.
  b) When the temperature is measured it is less than expected. Give 2 reasons for this.

a) mass is measured in kilograms.
250 grams = 0.25 kg

$$\frac{\text{change in}}{\text{temperature}} = \frac{\text{energy}}{\text{mass x SHC}}$$

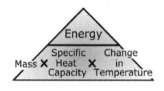

$$= \frac{5000 \text{ J}}{0.25 \text{ kg} \times 4{,}200 \text{ Jkg}^{-1}\text{K}^{-1}}$$

= 4.8 degrees Celsius

final temperature = initial temperature + temperature change
 = (35 + 4.8) degrees Celsius
 = 39.8 degrees Celsius

b) i) Energy is also used in heating the container (the glass).
  ii) Heat energy is lost heating the surroundings (the air).

6.6) I push a 10 kg box with a force of 200 N from one end of a corridor to the other end, 30 metres away. How much work have I done?

The weight of the object has been put in here but in this case it is unnecessary information. As we are not moving the object up or down the weight is irrelevant and the basic equation remains.

work done = force x distance
 = 200 N x 30 m
 = 6,000 J or 6 kJ

6.7) A cart requires a force of 300 N to move it. The cart is pushed for a while until the child pushing it gets tired. By the time the child has stopped to rest he has performed 3,750 J of work.
   a) How far has the cart been pushed
   b) The total time that the child spent pushing the cart was 1 minute. How much power did he generate?

a) distance = $\dfrac{\text{work done}}{\text{force}}$

$= \dfrac{3{,}750 \text{ J}}{300 \text{ N}}$

= 12.5 metres

b) power = $\dfrac{\text{work done}}{\text{time}}$

$= \dfrac{3{,}750 \text{ J}}{60 \text{ s}}$

= 62.5 J/s
= 62.5 watts

6.8) An elevator moves a 2,000 kg car 10 metres vertically upward.
   a) How much work is done by an elevator moving a 2,000 kg car 10 metres vertically upwards?
   b) The car is lifted the 10 metres in 40 seconds. What power does the lift generate?

a) In this case we are moving the object upwards. The force that is being generated is equal to the weight of the car.

weight = mass x gravity
= 2,000 kg x 10 ms^{-2}
= 20,000 N.

This is also the value of the force (as we are lifting it up we have to overcome the weight).

work done = force x distance
= 20,000 N x 10 metres
= 200,000 J

b) power = $\dfrac{\text{work done}}{\text{time}}$

$= \dfrac{200{,}000 \text{ J}}{40 \text{ s}}$

= 5,000 Js^{-1}
= 5 kW

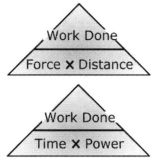

6.9) A train engine is producing power at a rate of 100 kW. This is how it is able to move forwards. The train is travelling at a constant speed. How much work is done over 1 minute by the train engine?

work done = time x power
= 60 s x 100,000 W
= 6,000,000 J
= 6 MJ

6.10) A car engine produces a force of 10 kN when it is moving at 25 ms^{-1}.
   a) What power does it generate?
   b) If 1 horsepower is 746 watts how many hp does the car generate?

Every second the car moves a distance of 25 m.

a) work done = force x distance
         = 10,000 N x 25 m
         = 250,000 J

power = $\dfrac{\text{work done}}{\text{time}}$

   = $\dfrac{250,000 \text{ J}}{1 \text{ s}}$

   = 250,000 W
   = 250 kW

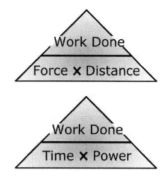

b) $\dfrac{\text{number of}}{\text{horsepower}}$ = $\dfrac{\text{total power}}{\text{power for 1 horsepower}}$

   = $\dfrac{250,000 \text{ W}}{746 \text{ W per hp}}$

   = 335

Or the answer could be written as 335 hp.

Bonus Questions 6.1: Work, Energy, Power and Heat

6.1.1) Calculate the energy required to raise the temperature of a material of mass 3 kg and specific heat capacity of 400 Jkg^{-1}K^{-1} by a temperature of 30 K.

6.1.2) How much energy is needed to raise the temperature of a 20 kg block of aluminium of specific heat capacity 0.9 kJkg^{-1}K^{-1} from 30 degrees Celsius to 55 degrees Celsius?

6.1.3) Water has a specific heat capacity of 4,200 Jkg^{-1}K^{-1}. How much would the temperature of a 15 kg bucket of water rise by if you add 300 kJ of energy from an electrical heating element? You can ignore the mass of the bucket.

6.1.4) 200 J of energy are added to an unknown liquid of mass 20 g. The temperature of the liquid changes from 295 K to 312 K. Calculate the specific heat capacity of the material.

6.1.5) 5,000 J of thermal energy is added to a mass of 600 g of an unknown material which is hidden in a box. There is a thermometer in the box which registers an initial temperature of 15 degrees Celsius. After the energy is added, the temperature of the material changes to 34 degrees Celsius. Calculate the value of the specific heat capacity of the material. Using the table below identify the most likely material.

Material	Specific Heat Capacity
Steel	452 Jkg^{-1}K^{-1}
Aluminium	900 Jkg^{-1}K^{-1}
Water	4.2 kJkg^{-1}K^{-1}

6.1.6) The engine temperature of a high performance car rises from 20 degrees Celsius to 120 degrees Celsius while running. The mass of the engine is 190 kg. The engine is made from steel. Calculate how much energy has been wasted heating up the engine.
Steel has a specific heat capacity of 0.452 kJkg^{-1}K^{-1}.

6.1.7) 1,500 J of thermal energy are needed to raise the temperature of a material with a specific heat capacity of 390 Jkg^{-1}K^{-1} by 30 K. What mass of the material is there?

6.1.8) A worker is pushing a box along a surface. The worker uses a force of 520 N and moves the box through a distance of 23 m. How much work has been done?

6.1.9) Which of the following does the most work?
   a) A car, average force = 600 N, distance travelled = 200 km
   b) A minivan, average force = 1,100 N, distance travelled = 120 km
   c) A truck, average force = 2,000 N, distance travelled = 50 km

6.1.10) The empire state building has a height of 381 m.
   a) Calculate the amount of work done by an 80 kg person carrying 100 g of ice cream to the top of the building using the stairs.
   b) There are about 800,000 J of energy in 100 g of ice cream. Calculate whether someone on a diet eating the ice cream will have done enough work to ensure that they can eat the ice cream without gaining weight.

6.1.11) If you hold a book at arm length you will find that your muscles get tired and it will soon become painful. Calculate how much work is done holding the book stationary (assume the book has a mass of 0.4 kg) without moving it and comment on the answer. (Why do your muscles become tired?)

6.1.12) A car is being pushed by kind pedestrians with a force of 300 N for a distance of 120 m.
   a) Calculate the work done in pushing the car.
   b) If there is a resistance force of 40 N how much work is done against the resistance force?
   c) How much of the work done was useful work done?

6.1.13) A worker calculates that in order to lift a concrete pillar with a weight of 30 kN to the floor of the building in which it is being installed that they will need to do 2.7 MJ of work. How high is the pillar being lifted?

6.1.14) 2,000 J of work is done in moving an object 32 metres. How much force was used in moving the object?

6.1.15) Work is done at a rate of 600 watts for a time period of 14 seconds. Calculate the total work that is done in this time.

6.1.16) What rate of power is required to do 3,000 J of work in a time period of 15 seconds?

6.1.17) What power would be required to lift an object of mass 20 kg a distance of 3 m in a time period of 4 s?

6.1.18) A lady of mass 65 kg is exercising by running up some stairs. She runs up a vertical height of 20 metres in a time of 25 seconds. Calculate her power.

6.1.19) How much time does it take for a 1.5 kW kettle to do an equal amount of work to the energy required to lift 100 kg up a height of 150 metres?

6.1.20) When you perform a push-up you lift your centre of mass about 15 cm.
   a) Calculate the work done by an adult of mass 85 kg in doing a single push-up.
   b) How many push-ups would they need to do in 1 second to equal the energy changed from electrical to heat by a 1.5 kW kettle in 1 second?

Bonus Questions 6.2: Work, Energy, Power and Heat

6.2.1) How much energy is required to increase the temperature of a substance with a mass of 0.7 kg and a specific heat capacity of 650 Jkg^{-1}K^{-1} by a temperature of 45 K?

6.2.2) Calculate the amount of energy that is needed to raise the temperature of a 15 kg block of steel of specific heat capacity 0.452 kJkg^{-1}K^{-1} from 23 degrees Celsius to 60 degrees Celsius.

6.2.3) Calculate the temperature rise of a 20 kg container of water when you add 2,000 kJ of energy from an electrical heating element. You can ignore the mass of the container. Water has a specific heat capacity of 4,200 Jkg^{-1}K^{-1}.

6.2.4) When I add 300 J of energy to an unknown solid with a mass of 15 g, the temperature changes from 302 K to 345 K. Find the specific heat capacity of the material.

6.2.5) Some students are trying to decide what mysterious material is located inside a sealed container. They are allowed to add energy to heat the material and the can check the temperature readings but they are not allowed to see inside the box. They are given a list of 3 possible materials that may be the identity of the mystery material. They are told that there is 500 g of the material. The students add 3,000 J of thermal energy to the mass in the box. There is a thermometer in the box which registers an initial temperature of 20 degrees Celsius. After the energy is added, the temperature of the material changes to 34 degrees Celsius.
   a) Calculate the value of the specific heat capacity of the material.
   b) Using the table below identify the most likely material.
   c) Why is your value not identical to any of the values listed?

Material	Specific Heat Capacity
Copper	385 Jkg^{-1}K^{-1}
Iron	444 Jkg^{-1}K^{-1}
Water	4.2 kJkg^{-1}K^{-1}

6.2.6) In the cold winter some people heat up their car in order to ensure that it will run well. The mass of a car engine is 150 kg and the engine is made from steel. How much energy is needed to heat up the car engine from -20 degrees Celsius to 80 degrees Celsius?
Steel has a specific heat capacity of 0.452 kJkg^{-1}K^{-1}.

6.2.7) If I add 3,200 J of thermal energy to a material with a specific heat capacity of 420 Jkg^{-1}K^{-1} by 45 K. What mass of the material is there?

6.2.8) A delivery driver is pushing a cart along the ground. The worker uses a force of 300 newtons and moves the cart through a distance of 23 metres. How much work has been done?

6.2.9) Which of the following does the most work?
    A cyclist, average force = 90 N, distance travelled = 50 km
    A motorbike, average force = 620 N, distance travelled = 12 km
    A car, average force = 1200 N, distance travelled = 7 km

6.2.10) For the end of year celebrations a student is carrying 30 water balloons to the roof garden in their school. The student weighs 54 kg and the roof garden is at a height of 5 meters above the ground. Assuming each balloon weighs 300 g, calculate the work done by the student.

6.2.11) A student is performing an experiment. They push hard on the walls of the school with a force of 200 N. After a minute they stop pushing, feeling tired.
   a) How much work was done?
   b) Why do they feel tired?

6.2.12) A cyclist has a puncture. They are pushing their bike back to their house. They push with a force of 100 N for a distance of 670 m.
   a) Calculate the work done in pushing the bike.
   b) If there is a resistance force of 60 N, how much work is done against the resistance force?
   c) How much of the work done was useful work done?

6.2.13) 30 kJ of energy are supplied in order to lift an object with a weight of 200 kN to a platform. How high is the object being lifted to rest on the platform?

6.2.14) I perform 500 J of work in moving an object 15 meters. How much force was used in moving the object?

6.2.15) How much work is done if it is being performed at a rate of 500 watts for a time period of 10 minutes?

6.2.16) How much power needs to be provided to perform 3,500 J of work in a time period of 28 seconds?

6.2.17) What power has to be supplied to lift a 35 kg object a distance of 12.5 m in a time period of 15 s?

6.2.18) Three students with an average mass 58 kg run up a flight of stair while being timed. They climb a vertical height of 15 m in an average time of 16 s.
   a) Calculate their average power output.
   b) What is the combined power output?

6.2.19) A 2 kW kettle is boiling a litre of water. It is heating the water from 30 degrees Celsius to 100 degrees Celsius. An electric motor is lifting a 30 kg mass. It will lift it through a height of 20 m. It provides 40 J of useful energy each second to the task. Which task is completed first?

6.2.20) The world record for pull ups is 43 in 60 seconds. Take the mass of the person performing the pull ups to be 60 kg and assume that they are lifting their body a distance of 30 cm.
   a) Calculate the work done in doing a single pull up.
   b) What was their average power in the 60 seconds?

7. Radioactivity

The Periodic Table Game

Let's play a game using the periodic table.

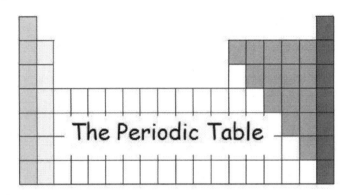

We're not going to use the complete periodic table for this explanation. This would work with the whole table but it is a little easier to focus on what is happening in a small part when there are fewer elements involved. You can find the complete periodic table just before the example questions.

The game is rather simple.
- Every time you read the word **alpha (α)** you move **down 2** numbers.
- Every time you read the word **beta (β)** you move **up 1** number
- Whenever you read the word **gamma (γ)** you **don't move**.
- Change the element name to match the new number.

Let's begin with an example:
Try this: - Place your finger on the element In (number 49).
- Follow the rules above using these steps: gamma, beta, alpha, alpha, beta, beta, beta.

					Al	Si
					13 Aluminium	14 Silicon
Ni	Cu	Zn	Ga	Ge
28 Nickel	29 Copper	30 Zinc	31 Gallium	32 Germanium
Pd	Ag	Cd	In	Sn
46 Palladium	47 Silver	48 Cadmium	49 Indium	50 Tin
Pt	Au	Hg	Tl	Pb
78 Platinum	79 Gold	80 Mercury	81 Thallium	82 Lead

Your finger should now be back on In (number 49).

The process: In (49) →γ gives→ In (49) →β gives→ Sn (50) →α gives→ Cd (48) →α gives→ Pd (46) →β gives→ Ag (47) →β gives→ Cd (48) →β gives→ In (49) - back at the start.

Let's try it again with a few questions

		Al 13 Aluminium	Si 14 Silicon	
Ni 28 Nickel	Cu 29 Copper	Zn 30 Zinc	Ga 31 Gallium	Ge 32 Germanium
Pd 46 Palladium	Ag 47 Silver	Cd 48 Cadmium	In 49 Indium	Sn 50 Tin
Pt 78 Platinum	Au 79 Gold	Hg 80 Mercury	Tl 81 Thallium	Pb 82 Lead

1) Start with Ge (number 32): alpha, gamma, alpha, beta, beta, alpha, gamma.

2) Start with Au (number 79): beta, beta, beta, gamma, alpha, beta, beta

3) Start with Hg (number 80): alpha, beta, beta, beta, gamma, gamma, gamma, alpha

Answers

1 – start with Ge (32)	2 – start with Au (79)	3 – start with Hg (80)
→ alpha - Zn (30)	→ beta - Hg (80)	→ alpha - Pt (78)
→ gamma - Zn (30)	→ beta - Ti (81)	→ beta - Au (79)
→ alpha - Ni (28)	→ beta - Pb (82)	→ beta - Hg (80)
→ beta - Cu (29)	→ gamma - Pb (82)	→ beta - Ti (81)
→ beta - Zn (30)	→ alpha - Hg (80)	→ gamma - Ti (81)
→ alpha - Ni (28)	→ beta - Ti (81)	→ gamma - Ti (81)
→ gamma - Ni (28)	→ beta - Pb (82)	→ gamma - Ti (81)
– end with Ni (28)	– end with Pb (82)	→ alpha - Au (79)
We changed germanium into nickel	We changed gold into lead	– end with Au (79) We changed mercury into gold

Now we're going to repeat the game. This time however we'll be using the number at the top and also using the number that we see at the bottom. Sometimes, depending on the book that you are using or the course that you are following, these numbers will be reversed. The best approach when this happens is to examine them in terms of the large number, called the atomic mass number or nucleon number (in this periodic table view at the top) and the smaller number, called the proton number or atomic number (in this periodic table view at the bottom).

Complete rules:
- Alpha (α) → small number -2 (move down 2 places on the table), big number -4.
- Beta (β) → small number +1 (move up 1 place on the table), big number nothing happens.
- Gamma (γ) → no change for small or big number. (Stay where you are on the table.)
- Element name → match this to the small number only.

	α	β	γ
big	-4	0	0
small	-2	+1	0

Let's begin with an example.

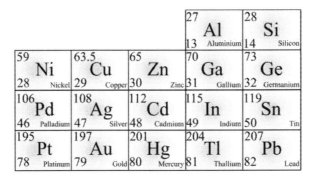

Place your finger on $^{65}_{30}Zn$: alpha, beta, gamma, gamma → You should now have $^{61}_{29}Cu$

The process: Start $^{65}_{30}Zn \xrightarrow{\alpha \text{ gives}} {}^{61}_{28}Ni \xrightarrow{\beta \text{ gives}} {}^{61}_{29}Cu$

→ and the 2 gammas (γ) have no effect. This leaves us with $^{61}_{29}Cu$

As the larger value (called the atomic mass number) for copper is different from the periodic table value we call it an **isotope** of copper.

Let's try a few questions with this.

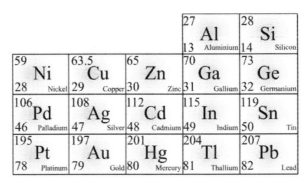

1) Start with $^{119}_{50}Sn$ then - gamma, gamma, alpha, alpha, beta

2) Start with $^{59}_{28}Ni$ then - beta, gamma, beta, alpha, gamma

3) Start with $^{207}_{82}Pb$ then - gamma, alpha, gamma, beta, alpha, gamma

Answers

1 – start with $^{119}_{50}$Sn

→ gamma $^{119}_{50}$Sn

→ gamma $^{119}_{50}$Sn

→ alpha $^{115}_{48}$Cd

→ alpha $^{111}_{46}$Pd

→ beta $^{111}_{47}$Ag

– end with $^{111}_{47}$Ag

This is an isotope of silver (Ag)

2 – start with $^{59}_{28}$Ni

→ beta $^{59}_{29}$Cu

→ gamma $^{59}_{29}$Cu

→ beta $^{59}_{30}$Zn

→ alpha $^{55}_{28}$Ni

→ gamma $^{55}_{28}$Ni

– end with $^{55}_{28}$Ni

This is an isotope of nickel (Ni)

3 – start with $^{207}_{82}$Pb

→ gamma $^{207}_{82}$Pb

→ alpha $^{203}_{80}$Hg

→ gamma $^{203}_{80}$Hg

→ beta $^{203}_{81}$Ti

→ alpha $^{199}_{79}$Au

→ gamma $^{199}_{79}$Au

– end with $^{199}_{79}$Au

We have turned lead into gold!

How it works: The small number tells us the number that the element has. It tells us how much positive charge it has in the nucleus. Each element has its own number. If this number is changed then we have actually changed the type of element that we have. This can be called the proton number or the atomic number and it is equal to the amount of positive charge. (This is important as later we will write an electron with a negative number as it has a negative charge.)

Alpha decay carries away 2 of these protons. As they are both positively charged they need another 2 particles without charge to help hold them together. These are neutrons. The number of positive particles taken away from the nucleus is 2, so the small number will drop by 2. The total number of particles taken away from the nucleus is 4 (2 protons + 2 neutrons) so the large number will drop by 4.

The large number is the total number of protons and neutrons. This is called the atomic mass number, the mass number or the nucleon number.

Beta decay removes only a small particle which is much lighter than a proton and is why the large number doesn't change. This small particle is an electron but it is made when a neutron breaks down into a proton and electron. The negative electron is thrown out (electrons belong outside the nucleus) leaving 1 more positive charge inside the nucleus. This means that the small number increases by 1.

Gamma decay takes away an amount of energy but no other 'stuff'. This is why nothing changes.

Writing the element:

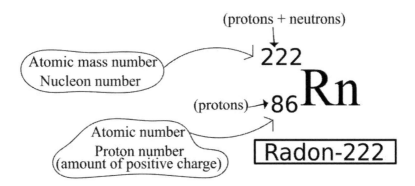

How do you remember whether which number goes on the top and bottom?

If you built a building like the element, it could fall over because it is top heavy.

The Radiation Story

Once upon a time I was walking home and I saw a big lead gem. I thought this was strange as lead does not form gems. I then realized I had to hold my jeans up as I had lost my aluminium belt. This was okay though as I found two pins and a paper hat sitting on the table when I went inside. The end.

<u>Lead gem 0utside</u>
<u>Lost aluminium bel</u>t
<u>Found 2 pi</u>ns and a <u>paper ha</u>t <u>inside</u>

<u>Lead g(em) 0utside</u>:
    <u>Lead</u> gem - **Lead** reduces gamma radiation.
    <u>g(em)</u> - **G**amma radiation is an **e-m** wave (e-m stands for electromagnetic).
    <u>0</u>utside - It has a charge of **0**.
    <u>outside</u> - It is the most dangerous type of radiation when **outside** the body.

<u>Lost aluminium bel</u>t:
    <u>Lost</u> - The belt has been lost. It has a charge of **-1**.
    <u>aluminium b</u>elt - A few mm of **aluminium** stops beta radiation
    <u>bel</u>t - **B**eta radiation is an **el**ectron.

<u>Found 2 pi</u>ns and a <u>paper ha</u>t <u>inside</u>
    <u>Found 2</u> - The items have been found – it has a charge of **+2**.
    2 <u>pi</u>ns – It is made of **2 p**rotons and **2 n**eutrons. This means that it has an atomic number of 2 and an
        atomic mass of 2 + 2 = 4. So there are 4 nucleons in an alpha particle.
    <u>paper</u> hat – A sheet of **paper** is enough to stop **a**lpha radiation.
    <u>ha</u>t - A **h**elium nucleus is an **a**lpha particle.
    <u>inside</u> - It is the most dangerous type of radiation when **inside** the body.

Alpha radiation: Helium nucleus

Charge: 2+

Alpha particles are made up of a lot of 2's – they have a charge of 2+, they hold 2 protons, they also hold 2 neutrons, giving a mass of 2 + 2 which is 4 atomic mass units or 4 u. This means that they are made of 4 nucleons (a nucleon is a particle in the nucleus. Both protons and neutrons are nucleons). An alpha particle is a helium nucleus.

Alpha radiation is the most ionising of the types of radiation. This means that it can cause a lot of cell damage which can cause cancer. It is most dangerous when it is inside the body because of this. It is least dangerous when it is outside the body as it is stopped very easily so it is least likely to penetrate the skin to cause damage.

Writing an alpha particle: $^{4}_{2}\alpha$ or $^{4}_{2}He$

Beta radiation: Electron

Charge: 1-

Beta radiation is an electron. It is also an elementary particle because it isn't made up of other things. It is not the most or least dangerous of any radiation type.

Writing a beta particle: $^{0}_{-1}\beta$ or $^{0}_{-1}e$

Gamma radiation: energy or EM ray

Charge: 0

Gamma radiation is pure energy. It is the most dangerous outside the body. It has no charge and is an EM ray.

Writing a gamma ray: $^{0}_{0}\gamma$

How is it Stopped?

Here is another way to remember how to stop the different types of radiation.

A penguin
beating **aluminium**
gets **lead**
by chance.

A penguin → **a**lpha radiation is stopped by **p**aper

beating **aluminium** → **b**eta radiation is stopped by a few mm of **aluminium**

gets **lead** → **g**amma radiation is reduced by a few cm of **lead**

By chance → radiation is unpredictable and random

Radioactive Decay Equations

Now that we have learned this we can examine a radioactive decay equation written completely.

This is an example of an alpha decay equation:

$$^{214}_{84}\text{Po} \longrightarrow {}^{210}_{82}\text{Pb} + {}^{4}_{2}\alpha$$

Let's examine the numbers on the left and right hand sides. For both the top and the bottom, the numbers on the right should be equal to the numbers on the left added together.

Examining the top gives us: $214 = 210 + 4$
Examining the bottom gives us: $84 = 82 + 2$

These numbers on the left and right have to be equal, and we already know the numbers for the alpha, beta and gamma radiation. We can find the value of any missing numbers using the above approach (this comes up often in questions). We will examine this after showing examples of the other types of radiation.

This is an example of a beta decay equation:

$$^{210}_{83}\text{Bi} \longrightarrow {}^{210}_{84}\text{Po} + {}^{0}_{-1}\beta$$

on the top: $210 = 210 + 0$
and on the base: $83 = 84 - 1$

This is an example of a gamma decay equation:

$$^{230}_{90}\text{Th}^{*} \longrightarrow {}^{230}_{90}\text{Th} + {}^{0}_{0}\gamma$$

(* means it is in an excited state)

The asterisk here shows that the nucleus has too much energy and is able to release a gamma ray. A gamma ray has no charge and no nucleons so examining the numbers gives:

on top: $230 = 230 + 0$
on base: $90 = 90 + 0$

One of the key concepts that we need to know here (the reason the numbers don't change) is that for all radioactive decay equations the total atomic number and the total mass number before and after do not change.

Let's examine a question using this:

Complete the radioactivity decay equation below.

$$^{234}_{\square}Th \longrightarrow ^{\square}_{91}Pa + ^{\square}_{\square}\beta + ^{\square}_{\square}\gamma$$

We can immediately write down the values of the top line for beta and gamma radiation. This will enable us to calculate the missing value from the symbol Pa.

234 = missing value + 0 + 0

This tells us that the missing top value for Pa is 234.

Examining the bottom values then gives

Missing value = 91 + (-1) + 0

So the missing value = 91 - 1 = 90

These numbers can then be added to give our final answer:

$$^{234}_{90}Th \longrightarrow ^{234}_{91}Pa + ^{0}_{-1}\beta + ^{0}_{0}\gamma$$

Subatomic Particles

There are 2 main types of fundamental subatomic particle; leptons and quarks.

Quarks

- Are what neutrons and protons are made up of. (Electrons are a type of fundamental particle so are not made of quarks.)
- They come in different 'flavours'.

So let's look at how we can remember what neutrons and protons are of.

They key information to remember is: - The charge
                                       - Whether it is an up or a down quark.

Quarks love pudding!

(quarks love puud-ddun)

(puud-ddun here would be pronounced as pudding!)

This tells what 3 quarks protons and neutrons are made from.

Protons: are made an **U**p, an **U**p and a **D**own quark

    (Protons - up, up, down, which is where the PUUD comes from.
        - You may also see it written UUD rather than up, up, down)

Neutrons: A **D**own, a **D**own and an **U**p quark make a **N**eutron

    (down, down, up - Neutron, which is where the DDUN comes from.
        - You may also see it written DDU rather than down, down, up.)

Therefore, to help us remember this, we need to know that quarks love puudddun!

There is one more thing you need to know about the Up and Down quarks. This is their charge.

To help you remember:

→ Up is positive, down is negative (like a graph).
→ Up is bigger than down.

You often ask people: 'What have you been **up** to?' (or **two**)

→ The **up** quarks have an electric charge of $+\frac{2}{3}$

→ So **down** quarks have the remaining charge $-\frac{1}{3}$

A little above this level:

Well it seems almost wrong to have radioactivity without ever having a mutant (or spelled wrong as a mutent). This is why you are going to learn a little bit more about subatomic particles. If you are running short on time you can just answer questions on the topics you need to know or you could stretch yourself further.

Now we can write - Bring out the mutant!

The **quark** was **strangely charming,**
From the **bottom** to the **top**.
It jumped **up**, it jumped **down**,
and it jumped all around,
while the **mu-ten**t jumped on the mop.

This tells us how to remember:

- The 3 different of **quark** pairs. These are: **strange** and **charm**, **bottom** and **top**, **up** and **down**.

It also tells us how to remember the different types of lepton. (A lepton is an elementary particle just as quarks are elementary particles. They cannot be broken into smaller pieces.)

- There are actually 6 types of leptons:
    1. **mu**on
    2. **t**au
    3. **e**lectron
    4. **m**uon **n**eutrino
    5. **t**au **n**eutrino
    6. **e**lectron **n**eutrino.

When you put them together you really do get a **mu-ten**t!

Examples: Radioactivity

Use the periodic table below to help you answer questions on alpha, beta and gamma radiation using the radiation game.

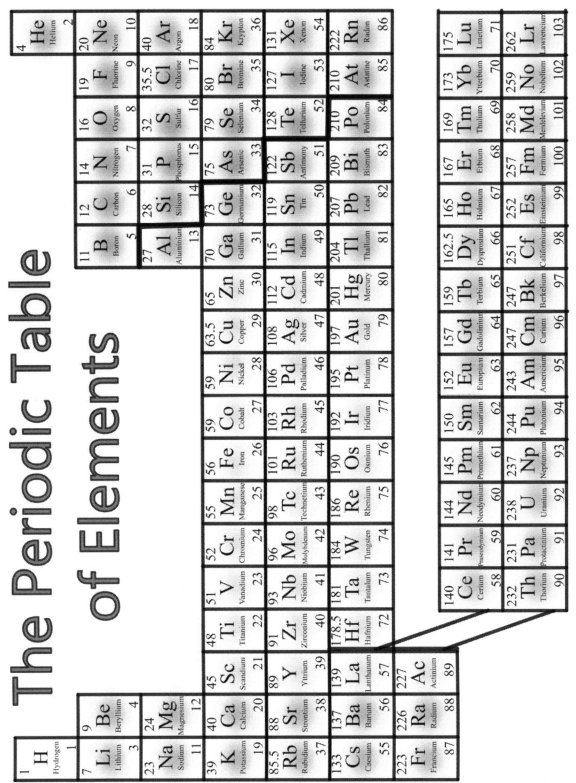

Use the periodic table on the previous page to answer the next three questions.

Example 1) What element do you end up with if Sb undergoes the following series of radioactive emissions?
alpha → beta → gamma → alpha → beta

$^{122}_{51}Sb \xrightarrow{\alpha \text{ gives}} {}^{118}_{49}In \xrightarrow{\beta \text{ gives}} {}^{118}_{50}Sn \xrightarrow{\gamma \text{ gives}} {}^{118}_{50}Sn$

	α	β	γ
big	-4	0	0
small	-2	+1	0

$\xrightarrow{\alpha \text{ gives}} {}^{114}_{48}Cd \xrightarrow{\beta \text{ gives}} {}^{114}_{49}In$ - This is an isotope of indium.

Example 2) What element do you end up with if a palladium nucleus gives out the following radioactivity?
beta → gamma → alpha → gamma → gamma

$^{106}_{46}Pd \xrightarrow{\beta \text{ gives}} {}^{106}_{47}Ag \xrightarrow{\gamma \text{ gives}} {}^{106}_{47}Ag \xrightarrow{\alpha \text{ gives}} {}^{102}_{45}Rh$

	α	β	γ
big	-4	0	0
small	-2	+1	0

$\xrightarrow{\gamma \text{ gives}} {}^{102}_{45}Rh \xrightarrow{\gamma \text{ gives}} {}^{102}_{45}Rh$ - This is an isotope of rhodium.

Example 3) What element do you end up with if Al undergoes the following series of radioactive decay?
beta → beta → beta → alpha

$^{27}_{13}Al \xrightarrow{\beta \text{ gives}} {}^{27}_{14}Si \xrightarrow{\beta \text{ gives}} {}^{27}_{15}P \xrightarrow{\beta \text{ gives}} {}^{27}_{16}S$

	α	β	γ
big	-4	0	0
small	-2	+1	0

$\xrightarrow{\alpha \text{ gives}} {}^{23}_{14}Si$ - This is an isotope of silicon.

Example 4) Describe an experiment to test whether a type of radiation being produced is alpha radiation.

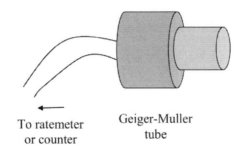

← To ratemeter or counter   Geiger-Muller tube

Step 1: Measure the background radiation over a period of time (about 1 minute). This is just the normal radiation that comes mostly from rocks and space. Some also comes from man-made sources like some hospital equipment and (about 1%) from nuclear testing.

149

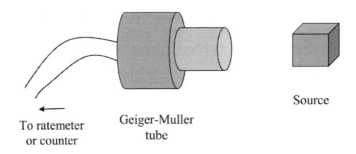

Step 2: Measure the initial activity of the radioactive source using a Geiger counter.

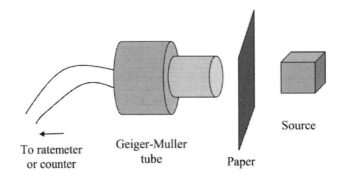

Step 3: Then place a piece of paper between the source and the Geiger counter. If the activity reduces to the level of the background radiation, then the radiation is alpha radiation because alpha radiation is stopped by a piece of paper.

Both gamma and beta radiation will pass through paper and the activity would be higher than the normal background radiation reading.

Example 5) What are the three different types of radiation and what are they made of?

Alpha radiation is made from 2 protons and 2 neutrons. It is a helium nucleus.

Beta radiation is an electron.

Gamma radiation is a high energy electromagnetic wave.

Example 6) State what is needed to stop the three different types of radioactivity.

Alpha radiation is stopped by a piece of paper.

Beta radiation is stopped by a few mm of aluminium.

Gamma radiation is stopped by a few cm of lead.

Example 7) I have 20 grams of a material, all of which is the same type of radioactive substance. After 20 minutes I still have 20 grams of material but 10 grams of radioactive substance remains, the rest having decayed.
   a) What is the half-life of the substance?
   How long will it take until I have:
   b) 1/4 of the original amount of radioactive substance
   c) 1/8 of the original amount of radioactive substance

a) Half-life = time it takes to reduce the amount of radioactive substance by 1/2

20 g x 1/2 = 10 grams

1/2 of the radioactive material decays in 20 minutes.

The radioactive half-life is 20 minutes.

b) 1/4 = 1/2 x 1/2

There will be 1/4 of the sample left after 2 half-lives.

1 half-life = 20 minutes,
2 half-lives = 40 minutes

c) 1/8 = 1/2 x 1/2 x 1/2

There will be 1/8 of the sample left after 3 half-lives.

3 half-lives = 20 minutes + 20 minutes + 20 minutes
             = 60 minutes

Example 8) A radioactive substance in a room has a half-life of 10 minutes and an initial activity of 200 Becquerels (Bq). It will be safe to enter the room once the activity has dropped below 50 Bq. Outside the room are a group of scientists. How long will they have to wait before they can safely enter the room?

200 Bq will change to 100 Bq after 1 half-life.
100 Bq will change to 50 Bq after 1 half-life.
Therefore, the activity will drop below 50 Bq after 2 half-lives.

1 half-life is 10 minutes.
2 half-lives is 20 minutes.
It will take 20 minutes until it is safe to enter the room.

Example 9) A radioactive substance has a half-life of 1 minute. If I start with 256 g of the radioactive substance draw an activity time graph to show the amount of activity for the first 8 minutes.

256 grams will change to 128 grams	1 half-life
128 grams will change to 64 grams	2 half-lives
64 grams will change to 32 grams	3 half-lives
32 grams will change to 16 grams	4 half-lives
16 grams will change to 8 grams	5 half-lives
8 grams will change to 4 grams	6 half-lives
4 grams will change to 2 grams	7 half-lives
2 grams will change to 1 grams	8 half-lives

Each half-life takes 1 minute so we can fill in a table:

Time / minutes	Amount left /grams
0	256
1	128
2	64
3	32
4	16
5	8
6	4
7	2
8	1

We can draw a graph with this information.

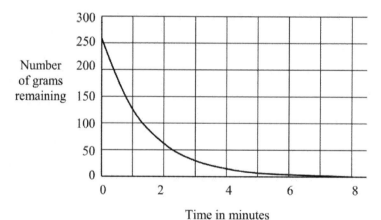

Time in minutes

Example 10) List the most dangerous types of radiation inside and outside the body in order of most dangerous to least dangerous.

Inside the body
Most dangerous is alpha radiation, then beta radiation and the least dangerous is gamma radiation.

Outside the body
Most dangerous is gamma radiation, then beta radiation and the least dangerous is alpha radiation.

Example 11) What are the dangers to humans of exposure to radiation and radioactive materials?

Radiation can cause: Radiation sickness
Cancer
It can damage DNA
It can damage cells
Death

Example 12) Describe the advantages and drawbacks of nuclear power stations.

Advantages:
They have a large energy density (A small amount of fuel can create a lot of electricity).
They do not produce greenhouse gases (after they have been built).

Disadvantages:
They are expensive to maintain.
It is difficult to dispose of the radioactive waste.

Exposure to radioactive waste can cause:
Radiation sickness
Cancer
It can damage DNA
It can damage cells

Example 13) Fill in the blanks to complete the radioactive decay equation below.

$$^{240}_{92}U \longrightarrow ^{\square}_{\square}\square + ^{\square}_{\square}\beta$$

Step 1:

You know that the mass number (top) for beta decay is always 0 and the proton number (bottom) is always -1.

Step 2: Fill in the top line. $240 = x + 0$
                                        $x = 240$

Step 3: Fill in the bottom line. $92 = x + (-1)$
                                        $x = 93$

Step 4: Use the periodic table to find the element name. This is the element that matches the proton number (the small number).

$$^{240}_{92}U \longrightarrow ^{240}_{93}Np + ^{0}_{-1}\beta$$

Example 14) Complete the radioactivity decay equation below by filling in the blanks.

$$^{244}_{\square}\square \longrightarrow ^{\square}_{\square}Cm + ^{\square}_{\square}\alpha + ^{\square}_{\square}\gamma$$

Step 1:

You know that the mass number (top) for alpha decay is always +4 and the proton number (bottom) is always +2. You also know that the mass and proton numbers for gamma are both zero.

Step 2: Fill in the top line. $\quad\quad 244 = x + 4 + 0$
$\quad\quad\quad\quad\quad\quad\quad\quad\quad\quad\quad\quad x = 240$

Step 3: Find the proton number (bottom) for Cm from the periodic table. It is 96.

Step 4: Fill in the bottom line. $\quad\quad x = 96 + 2 + 0$
$\quad\quad\quad\quad\quad\quad\quad\quad\quad\quad\quad\quad = 98$

Step 4: Use the periodic table to find the element that has a proton number of 94. This is Pu (plutonium).

$$^{244}_{98}Cf \longrightarrow ^{240}_{96}Cm + ^{4}_{2}\alpha + ^{0}_{0}\gamma$$

Questions: Radioactivity

7.1) What element do you end up with if lead undergoes the following series of radioactive emissions?
alpha → alpha → gamma → alpha

7.2) What element do you end up with if an In nucleus gives out the following radiation?
beta → beta → beta → alpha

7.3) What element do you end up with if Ne undergoes the following series of radioactive decay?
alpha → gamma → gamma → beta

7.4) What are the following types of radiation made from and what type of charge do they possess?
   a) Alpha radiation
   b) Gamma rays
   c) Beta radiation

7.5) What is required to stop the following types of radiation?
   a) Gamma
   b) Beta
   c) Alpha

7.6) Describe an experiment to tell if a type of radiation being released by a radioactive source is beta radiation.

7.7) Describe an experiment to tell if a type of radiation being released by a radioactive source is gamma radiation.

7.8) I have 800 grams of radioactive material. After 1 hour I have 50 grams of radioactive material remaining. The other 750 grams has decayed into a non-radioactive material. What is the half-life of the radioactive material?

7.9) Describe an experiment to measure the radioactive decay of a material.

7.10) Describe the drawbacks of nuclear power.

7.11) Fill in the blanks below to complete the radioactive decay equation.

$$^{\square}_{\square}Po \longrightarrow {}^{206}_{83}\square + {}^{\square}_{\square}\alpha + {}^{\square}_{\square}e$$

7.12) Complete the radioactivity decay equation by filling in the blanks below.

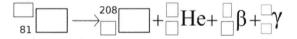

$$^{\square}_{81}\square \longrightarrow {}^{208}_{\square}\square + {}^{\square}_{\square}He + {}^{\square}_{\square}\beta + {}^{\square}_{\square}\gamma$$

Answers: Radioactivity

7.1) What element do you end up with if lead undergoes the following series of radioactive emissions?
alpha → alpha → gamma → alpha

I can find the answer to this question by following the rules to the periodic table game and using the periodic table at the beginning of the examples section.

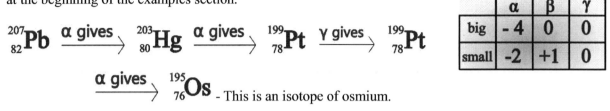

$^{207}_{82}Pb \xrightarrow{\alpha \text{ gives}} {}^{203}_{80}Hg \xrightarrow{\alpha \text{ gives}} {}^{199}_{78}Pt \xrightarrow{\gamma \text{ gives}} {}^{199}_{78}Pt$

$\xrightarrow{\alpha \text{ gives}} {}^{195}_{76}Os$ - This is an isotope of osmium.

	α	β	γ
big	-4	0	0
small	-2	+1	0

7.2) What element do you end up with if an In nucleus gives out the following radiation?
beta → beta → beta → alpha

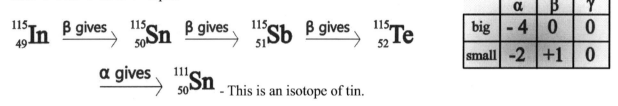

$^{115}_{49}In \xrightarrow{\beta \text{ gives}} {}^{115}_{50}Sn \xrightarrow{\beta \text{ gives}} {}^{115}_{51}Sb \xrightarrow{\beta \text{ gives}} {}^{115}_{52}Te$

$\xrightarrow{\alpha \text{ gives}} {}^{111}_{50}Sn$ - This is an isotope of tin.

	α	β	γ
big	-4	0	0
small	-2	+1	0

7.3) What element do you end up with if Ne undergoes the following series of radioactive decay?
alpha → gamma → gamma → beta

$^{20}_{10}Ne \xrightarrow{\alpha \text{ gives}} {}^{16}_{8}O \xrightarrow{\gamma \text{ gives}} {}^{16}_{8}O \xrightarrow{\gamma \text{ gives}} {}^{16}_{8}O$

$\xrightarrow{\beta \text{ gives}} {}^{16}_{9}F$ - This is an isotope of fluorine.

	α	β	γ
big	-4	0	0
small	-2	+1	0

7.4) What are the following types of radiation made from and what type of charge do they possess?

a) Alpha radiation:
      Alpha radiation is made of 2 protons and 2 neutrons.
      It is a helium nucleus.
      It has a charge of 2 positive.

b) Gamma rays:
      Gamma rays are high energy electromagnetic waves.
      They have no charge.

c) Beta radiation:
      Beta radiation is made up of electrons.
      It has a charge of 1 negative.

7.5) What is required to stop the following types of radiation?

a) Gamma: Gamma radiation is reduced by several centimetres of lead.

b) Beta: Beta radiation is stopped by a few millimetres of aluminium.

c) Alpha: Alpha radiation is stopped by a thin piece of paper.

7.6) Describe an experiment to tell if a type of radiation being released by a radioactive source is beta radiation.

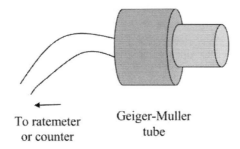

Step 1: Measure the background radiation. This is done by taking a reading over a time of 1 minute with no radioactive source present.

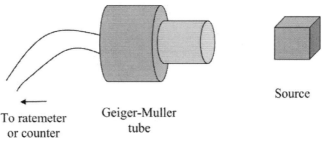

Step 2: Measure the activity of the radioactive source using a Geiger counter.

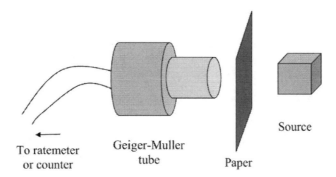

Step 3: Then place a piece of paper between the source and the Geiger counter. If the activity reduces to the level of the background radiation, then the radioactive emissions are being stopped by the paper. This tells us that the radiation released by the source is alpha radiation because alpha radiation is stopped by a piece of paper.

Both gamma and beta radiation will pass through paper and the activity would be higher than the normal background radiation reading.

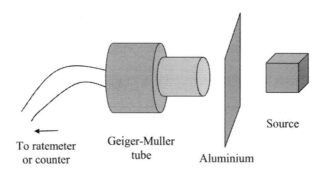

Step 4: Replace the piece of paper with a sheet of aluminium a few mm thick. This will stop the beta radiation. If the activity reduces to that of the background level with the aluminium but not with the paper, then you have just shown that the radiation being released is beta radiation.

7.7) Describe an experiment to tell if a type of radiation being released by a radioactive source is gamma radiation.

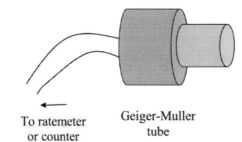

Step 1: Measure the background radiation for about a minute or so to get an average value.

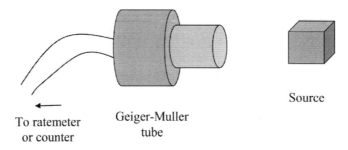

Step 2: Measure the initial activity of the radioactive source using a Geiger counter.

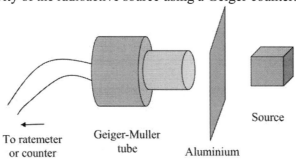

Step 3: Place a sheet of aluminium a few mm thick in between the source and the Geiger counter. If the radioactivity is not reduced, then the radiation from the source must be gamma radiation. This is the only type of radiation that can pass through a few mm of metal easily.

7.8) I have 800 grams of radioactive material. After 1 hour I have 50 grams of radioactive material remaining. The other 750 grams has decayed into a non-radioactive material. What is the half-life of the radioactive material?

800 grams to 400 grams = 1 half-life
400 grams to 200 grams = 1 half-life
200 grams to 100 grams = 1 half-life
100 grams to 50 grams = 1 half-life

Total number of half-lives = 4

Total amount of time = 1 hour (60 minutes)

Time for 1 half-life = 60 minutes / 4

1 half-life = 15 minutes

7.9) Describe an experiment to measure the radioactive decay of a material.

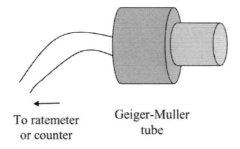

To ratemeter or counter   Geiger-Muller tube

Step 1: Measure the background radiation (this is just the normal radiation that comes mostly from rocks and space. Some also comes from man-made sources like some hospital equipment and (about 1%) from nuclear testing) over a period of time (about 1 minute).

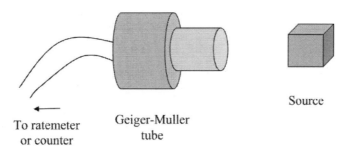

To ratemeter or counter   Geiger-Muller tube   Source

Step 2: Measure the initial activity of the radioactive source using a Geiger counter. Take this value and subtract the value of the background radiation that you have already measured.

Step 3: Continue measuring the activity of the source until its activity has dropped to ½ of the original value.

The time when the activity – the background activity = ½ the original activity of the source is the half-life of the radioactive source.

7.10) Describe the drawbacks of nuclear power:

It is expensive to maintain nuclear power stations.

It is difficult to dispose of the radioactive waste.

Accidents can result in an explosion with the release of radioactive materials.

Exposure to radioactive waste can:

    Cause radiation sickness

    Cause cancer

    Damage DNA

    Damage cells

7.11) Fill in the blanks below to complete the radioactive decay equation.

$$_{\square}^{\square}Po \longrightarrow {}_{83}^{206}\square + {}_{\square}^{\square}\alpha + {}_{\square}^{\square}e$$

Step 1: You know that alpha decay always has a mass number (top) of +4 and proton number (bottom) of +2. You also know that the 'e' stands for beta decay, which is just an electron. This means that the mass number is zero and the proton number is -1.

Step 2: Fill in the top line. $x = 206 + 4 + 0$
                                                     $= 210$

Step 3: Fill in the bottom line by using the periodic table to find the mass number for Po (polonium). This is 84.

Step 4: Find the element that has a mass number of 83. This is bismuth (Bi).

$$_{84}^{210}Po \longrightarrow {}_{83}^{206}Bi + {}_{2}^{4}\alpha + {}_{-1}^{0}e$$

7.12) Complete the radioactivity decay equation by filling in the blanks below.

$$^{\square}_{83}\square \longrightarrow ^{208}_{\square}\square + ^{\square}_{\square}He + ^{\square}_{\square}\beta + ^{\square}_{\square}\gamma$$

Step 1: You know that alpha decay is just a helium nucleus (He) so the mass number (top) is +4 and the proton number (bottom) is +2.

You also know that the mass number for beta decay is 0 and the proton number is -1 as beta decay is just an electron.

Finally, you know that gamma decay has no mass or proton number as it is just an electromagnetic wave.

Step 2: Fill in the top line.
$$x = 208 + 4 + 0 + 0$$
$$= 212$$

Step 3: Fill in the bottom line.
$$83 = x + 2 + (-1) + 0$$
$$x = 82$$

Step 4: Use the periodic table to find the element that matches the two known proton numbers.
 Proton number 83 is Bi (bismuth).
 Proton number 82 is Pb (lead).

$$^{212}_{83}Bi \longrightarrow ^{208}_{82}Pb + ^{4}_{2}He + ^{0}_{-1}\beta + ^{0}_{0}\gamma$$

Bonus Questions 7.1: Radioactivity

7.1.1) What element do you end up with if carbon undergoes the following series of radioactive emissions?
gamma → alpha → gamma → beta

7.1.2) What element do you end up with if a nickel nucleus gives out the following radiation?
gamma → alpha → beta → gamma

7.1.3) What element do you end up with if krypton undergoes the following series of radioactive decay?
beta → beta → gamma → beta

7.1.4) Describe an experiment that would allow someone to verify that a particular radioactive element only emitted alpha and gamma radiation.

7.1.5) Describe an experiment to verify that a particular radioactive element only emits alpha and beta radiation.

7.1.6) Describe an experiment to verify that a radioactive element only emits gamma and beta radiation.

7.1.7) A Geiger counter is set up to detect radiation and a sample is then placed in front of it. After 1 minute the Geiger counter has detected 34 counts of radiation. How many counts of radiation are there per second from the sample?

7.1.8) Why is it important when testing for radioactivity to check the background radiation count?

7.1.9) A source of the radioactive isotope astatine-210 has been left on a desk in a room. The room has been sealed. The initial radioactive count was 300 Bq. It will be safe to enter the room once the count has dropped below 40 Bq. The half-life of Astatine-210 is 8 hours 30 minutes. How long will it take before it is safe to enter the room again?

7.1.10) Radioactivity was discovered when a researcher left a sample of radium salts on top of a key which was in turn on top of an undeveloped photographic film plate. The plate was wrapped up but the radioactivity was able to penetrate through the wrapping and it created a picture of the key on the plate. Was the radioactivity being detected alpha radiation? Why or Why not?

7.1.11) Complete the following table to show the expected radioactivity count as the number of half-lives changes.

Number of Half-lives	Radioactive count /Bq
0	
1	1 600 000
2	
3	
4	
5	
6	
7	
8	

7.1.12) A radioactive medical isotope nitrogen-13 has a half-life of 10 minutes. Because of this it is normally made in the hospitals where it is used for PET scans. The radioactivity of a nitrogen-13 source is 400 Bq but it must be used before the radioactivity is less than 50 Bq. It will take 23 minutes to prepare the patient for the procedure, is there enough time for the patient to be prepared to undergo the procedure before the radioactivity is too low?

For the following four questions, complete the boxes for the radioactive decay equations.

7.1.13) $^{238}_{\square}U \longrightarrow {}^{\square}_{90}Th + {}^{\square}_{\square}\alpha$

7.1.14) $^{219}_{85}At^* \longrightarrow {}^{\square}_{\square}At + {}^{\square}_{\square}\gamma$

7.1.15) $^{14}_{6}C \longrightarrow {}^{\square}_{\square}N + {}^{\square}_{\square}\beta$

7.1.16) An astatine nucleus undergoes 2 alpha decays and a beta decay. Complete the following decay equation.

$$^{219}_{85}At \longrightarrow {}^{\square}_{\square}Pb + 2{}^{\square}_{\square}\alpha + {}^{\square}_{\square}e$$

7.1.17) What are the processes that occur in the sun to produce energy?

7.1.18) Describe the main difference between nuclear fission and nuclear fusion

7.1.19) A radioactive iodine tracer may be ingested in a hospital in order to allow the hospital staff to monitor a patient's thyroid. What type(s) of radiation do you think that the iodine will release and why?

7.1.20) List 3 benefits of nuclear power.

7.1.21) List 3 dangers of nuclear power.

Bonus Questions 7.2: Radioactivity

7.2.1) What element do you end up with if Nb undergoes the following series of radioactive emissions?
gamma → alpha → gamma → beta

7.2.2) What element do you end up with if a magnesium nucleus gives out the following radiation?
beta → gamma → alpha → gamma

7.2.3) What element do you end up with if vanadium undergoes the following series of radioactive decay?
beta → beta → gamma → beta

7.2.4) Explain why alpha radiation is more dangerous than gamma radiation inside the body.

7.2.5) Explain why gamma radiation is more dangerous than alpha radiation outside the body.

7.2.6) Describe an experiment to verify that a radioactive element only emits beta radiation.

7.2.7) An experiment is set up so that a Geiger counter is able to detect and count radiation. A background reading is then taken and the value for the background radiation is 15 counts per minute. A sample is then placed in front the counter. After 1 minute the Geiger counter has detected 34 counts of radiation. How many counts of radiation are there per second from the sample?

7.2.8) List some possible sources of background radiation.

7.2.9) A fraction of a gram of plutonium has been left in the corner of a room. It will be safe to enter the room again once the plutonium activity has dropped to $1/1024^{th}$ of its current activity. The half-life of plutonium is 24,000 years. When will it be safe to enter the room again?

7.2.10) In the 1950's some people thought that it might be a good idea to power planes with nuclear reactors, is this a good idea? Why or Why not?

7.2.11) Complete the following table to show the expected radioactivity count as the number of half-lives changes.

Number of Half-lives	Radioactive count /Bq
0	
1	
2	
3	
4	500, 000
5	
6	
7	
8	

7.2.12) How much material is needed to stop gamma radiation?

For the following questions, fill in the boxes for the radioactive decay equations:

7.2.13) $^{238}_{92}\square \rightarrow ^{\square}_{\square}\square + ^{\square}_{\square}e$

7.2.14) $^{\square}_{\square}Ra \rightarrow ^{224}_{\square}\square + ^{\square}_{\square}\alpha + ^{\square}_{\square}\gamma$

7.2.15) $^{\square}_{\square}\square \rightarrow ^{196}_{\square}Au + ^{\square}_{\square}He + ^{\square}_{\square}\gamma$

7.2.16) $^{112}_{\square}\square \rightarrow ^{\square}_{48}\square + ^{\square}_{\square}\gamma$

7.2.17) Give 2 reasons why nuclear power may be preferable to coal power.

7.2.18) Name the process by which;
  a) small nuclei are merged together releasing energy.
  b) large nuclei are broken apart releasing energy.

7.2.19) Using the table below, decide which of the isotopes is most suited to being injected into the body to act as a tracer for medical purposes.

Name of Isotope	Radiation Released
Polonium	Alpha
Iodine	Gamma
Carbon	Beta

7.2.20) Using the list below place the isotopes in order of danger outside the body from the safest to most dangerous.

Name of Isotope	Radiation Released
Polonium	Alpha
Iodine	Gamma
Carbon	Beta

7.2.21) List 3 dangers of nuclear power.

8. Charge, Energy, Efficiency and Electrical Power 1

Rabbits and Carrots

<u>V</u>ery <u>i</u>mportant <u>r</u>abbits
<u>V</u>ery **<u>large</u>** <u>r</u>abbits
<u>A</u>lso **<u>small</u>** <u>r</u>abbits

<u>V</u>ery	V	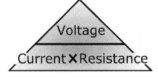
<u>i</u>mportant <u>r</u>abbits	I x R	
<u>V</u>ery **<u>large</u>** <u>r</u>abbits		<u>V</u>oltmeters have **<u>large</u>** resistance
<u>A</u>lso **<u>small</u>** <u>r</u>abbits		<u>A</u>mmeters have **<u>small</u>** resistance

Everybody knows that rabbits eat carrots!

<u>C</u>an't <u>t</u>aste <u>c</u>arrots
<u>e</u>ating <u>c</u>anned <u>v</u>egetables.
<u>E</u>at <u>t</u>hem <u>p</u>ickled.

<u>C</u>an't	C	
<u>t</u>aste <u>c</u>arrots	t x c	
	(Also Q = t x I)	
<u>e</u>ating	E	
<u>c</u>anned <u>v</u>egetables.	C x V	
<u>E</u>at	E	
<u>t</u>hem <u>p</u>ickled.	t x P	

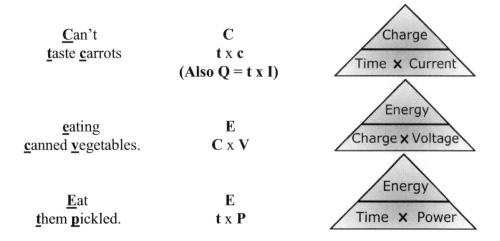

Electric Toothpaste

Efficiency is another key concept:

Use **e**lectric **t**oothpaste

Use   Useful
electric **t**oothpaste   e x t

Efficiency can only be between 1 and 0. A value of 1 represents 100% efficiency while 0 represents all of the energy being wasted, so 0% efficiency.

We can take whatever value we calculate for efficiency and multiply it by 100% to get the percentage efficiency. For example; an efficiency of 0.2 would give us a percentage efficiency of 0.2 x 100% = 20%.

Examples: Charge, Energy, Efficiency and Electrical Power 1

Example 1) Calculate the current in a circuit if the voltage supplied to the circuit is 12 volts and the resistance of the circuit is 3 ohms.

current = $\dfrac{\text{voltage}}{\text{resistance}}$

= $\dfrac{12 \text{ V}}{3 \text{ Ω}}$

= 4 amps

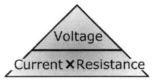

Example 2) The current in a circuit is 12 A and the resistance is 10 ohms. Calculate the voltage supplied to the circuit.

voltage = current x resistance
= 12 A x 10 Ω
= 120 volts

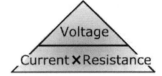

Example 3) The voltage supplied to a circuit is 20 V and the current in the circuit is 9 A. Calculate the resistance of the circuit.

resistance = $\dfrac{\text{voltage}}{\text{current}}$

= $\dfrac{20 \text{ V}}{9 \text{ A}}$

= 2.22 Ω

Example 4) A current of 3 amps travels past a point in a circuit for a time of 7 seconds. How much charge has moved past the point?

charge = time x current
= 7 s x 3 A
= 21 coulombs

Example 5) How long will it take a current of 10 amps to move a charge of 30 coulombs past a point in a circuit?

time = $\dfrac{\text{charge}}{\text{current}}$

= $\dfrac{30 \text{ C}}{10 \text{ A}}$

= 3 seconds

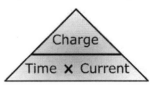

Example 6) I supply 30 J of energy to move a charge of 20 coulombs. What is the voltage supplied?

voltage = energy / charge

= 30 J / 20 C

= 1.5 volts

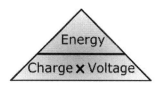

Example 7) 3 volts is applied to a circuit while a charge of 30 coulombs is moved through it. How much energy is needed for this process to occur?

1 volt = 1 joule per coulomb
energy = charge x voltage
= 30 coulombs x 3 volts
= 90 joules

Example 8) How much energy is needed to supply a 20 watt light for 30 seconds?

energy = time x power
= 30 s x 20 W
= 600 J

Example 9) A circuit is supplied with a total of 40 J of energy for 10 seconds. What power is dissipated in the circuit?

power = energy / time

= 40 J / 10 s

= 4 watts

Example 10) A 3 kW heater is turned on for an hour.
a) Calculate the total energy that is converted in this time.

energy = time x power
= 60 min x 60 seconds x 3,000 watts
= 10,800,000 J
= 10.8 MJ

b) Calculate this in kWh units.

energy = time x power
= 1 hour x 3 kW
= 3 kWh

Example 11) 3 light bulbs, each of power 60 watts, are turned on for 4 hours a day for 365 days (1 year).
a) How many kWh units have been used?

energy usage of each light bulb = 60 W
= 0.06 kW
energy usage of 3 light bulbs = 3 x 60 W = 180 W
= 0.18 kW
energy = time x power
energy = 4 hours x 365 days x 0.18 kW
= 263 kWh

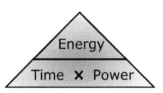

b) If each kWh costs 20p what is the total cost of the light bulbs for the year?

cost = 263 kWh x 20p
= 5,260p
= £52.60

Example 12) The light bulbs in the previous question were replaced with energy saving bulbs that each use 12 W.
a) What is the total energy saving for the year?

energy usage of each light bulb = 12 W
= 0.012 kW
energy usage of 3 light bulbs = 3 x 12 W = 36 W
= 0.036 kW
energy = time x power
energy = 4 hours x 365 days x 0.036 kW
= 52.6 kWh
cost = 52.6 kWh x 20p
= 1,051p
= £10.51

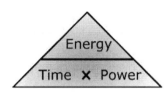

b) How much money is saved?
saving = cost before - cost after
= £52.60 - £10.51
= £42.09

Example 13) I use a 150 watt light bulb for 1 hour. How long would I have to use a 25 W energy saving light bulb before it would use the same amount of electricity?

energy used by 150 W bulb = energy used by 25 W bulb
time 1 x power 1 = time 2 x power 2

150 watts = 0.15 kW
25 watts = 0.025 kW

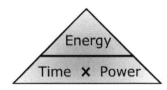

1 hour x 0.15 kW = time 2 x 0.025 kW

time 2 = $\dfrac{0.15 \text{ kWh}}{0.025 \text{ kW}}$

= 6 hours.

Example 14) An electric car is operating at 90% efficiency. If the electric car uses 400 kJ of energy to travel 1 km, calculate:
   a) Useful energy converted
   b) Energy wasted

a) 'Useful energy converted' is another way of saying 'useful work done'.

useful work done = efficiency x total energy in
            = 0.9 x 400 kJ
            = 360 kJ

b) energy wasted = total energy in – useful work done
            = 400 kJ – 360 kJ
            = 40 kJ

Example 15) Electricity is transported through overhead electrical wires. It is transported at an efficiency of 0.995 and 139.3 kW of energy is successfully transported;
   a) How much electrical energy is being transported in total?
   b) How much energy is being lost in transportation?

a)
total energy in = $\dfrac{\text{useful work done}}{\text{efficiency}}$

= $\dfrac{139.3 \text{ kW}}{0.995}$

= 140 kW

b) energy lost = total energy in – useful work done
            = 140 kW – 139.3 kW
            = 0.7 kW or 700 W

Example 16) A large engine is being used to lift a weight of 4,000 N to a height of 65 m. If the engine is only 35% efficient, how much weight could it lift if it was to be 100 % efficient?

(remember that weight = mass x gravity)

GPE = height x mass x gravity
    = height x weight
    = 65 m x 4,000 N
    = 260,000 J

Lifting 4,000 N to a height of 65 m takes 260 kJ of energy.

At 35% efficient:

total energy in = $\dfrac{\text{useful work done}}{\text{efficiency}}$

= $\dfrac{260 \text{ kJ}}{0.35}$

= 743 kJ

At 100% efficient:

GPE = height x mass x gravity
    = height x weight

weight = $\dfrac{\text{GPE}}{\text{height}}$

= $\dfrac{743,000 \text{ J}}{65 \text{ m}}$

= 11,430 N or 11.4 kN

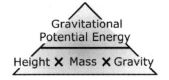

Questions: Charge, Energy, Efficiency and Electrical Power 1

8.1) A current of 12 amps flows around a circuit. How much time does it take before a charge of 50 C passes a point on the circuit?

8.2) 600 C of charge moves around a circuit in 2 minutes. What is the current in the circuit?

8.3) 12 kC of charge moves around a circuit in a day. What was the average amount of current that flowed in the circuit during that time period?

8.4) 12 J of energy are used to move a charge of 600 C. Calculate the voltage supplied.

8.5) 300 volts moves a charge of 40 C around a circuit. How much energy is changed in the process?

8.6) 70 J of energy is used to move a charge of 40 C around a circuit. Calculate the voltage supplied to the circuit.

8.7) In a house the energy usage is 10 kWh per day. If each kWh costs 20p calculate the yearly electricity bill.

8.8) Calculate how much money is saved per day when a family change from using eight 60 watt light bulbs for 5 hours a day to using eight 12 W energy saving bulbs. Each kWh costs 20p.

8.9) A 2 kW kettle takes 10 minutes to boil. Every day it is boiled 3 times to make cups of coffee.
   a) Calculate the number of kilowatt hours that are required for the kettle to boil every year.
   b) Calculate the cost if each kWh is 20p.

8.10) For the previous question calculate the amount of energy used, in joules, to make cups of coffee per year.

8.11) A power station operates at 45% efficiency, generating an output of 100 MW. Calculate the power available from the fuel that is put into the power station.

8.12) A system of power generation has an efficiency of 34%. This system generates 500 MW of energy to transport.
   a) How much energy is lost in the generation process?
   b) How much energy would be generated if the efficiency was to rise to 37%?

Answers: Charge, Energy, Efficiency and Electrical Power 1

8.1) A current of 12 amps flows around a circuit. How much time does it take before a charge of 50 C passes a point on the circuit?

time $= \dfrac{\text{charge}}{\text{current}}$

$= \dfrac{50 \text{ C}}{12 \text{ A}}$

$= 4.17$ s

8.2) 600 C of charge moves around a circuit in 2 minutes. What is the current in the circuit?

current $= \dfrac{\text{charge}}{\text{time}}$

total time (in seconds) = 2 minutes x 60 seconds

current $= \dfrac{600 \text{ C}}{120 \text{ s}}$

$= 5 \text{ Cs}^{-1}$
$= 5$ amps

8.3) 12 kC of charge moves around a circuit in a day. What was the average amount of current that flowed in the circuit during that time period?

current $= \dfrac{\text{charge}}{\text{time}}$

total time = 60 seconds in a minute x 60 minutes per hour x 24 hours in a day
$= 86{,}400$ s

current $= \dfrac{12{,}000 \text{ C}}{86{,}400 \text{ s}}$

average current $= 0.139$ amps

8.4) 12 J of energy are used to move a charge of 600 C. Calculate the voltage supplied.

voltage $= \dfrac{\text{energy}}{\text{charge}}$

$= \dfrac{12 \text{ J}}{600 \text{ C}}$

$= 0.02$ V
$= 20$ mV

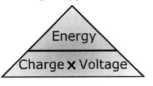

8.5) 300 volts moves a charge of 40 C around a circuit. How much energy is changed in the process?

energy = charge x voltage
       = 40 C x 300 V
       = 12,000 J
       = 12 kJ

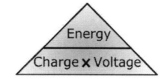

8.6) 70 J of energy is used to move a charge of 40 C around a circuit. Calculate the voltage supplied to the circuit.

voltage = $\dfrac{\text{energy}}{\text{charge}}$

      = $\dfrac{70 \text{ J}}{40 \text{ C}}$

      = 1.75 volts

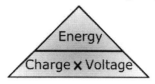

8.7) In a house the energy usage is 10 kWh per day. If each kWh costs 20p calculate the yearly electricity bill.

number of kWh per day x number of days = total number of kWh
             10 kWh x 365 days = 3,650 kWh per year

number of kWh used x cost of each kWh = total cost
total cost = 3,650 kWh x 20p
         = 73,000p
         = £730

8.8) Calculate how much money is saved per day when a family change from using eight 60 watt light bulbs for 5 hours a day to using eight 12 W energy saving bulbs. Each kWh costs 20p.

First calculate the cost of the eight 60 W bulbs.

amount of kW being used = amount of kW per bulb x 8 bulbs
      0.06 kW x 8 bulbs = 0.48 kW

number of kW hours = the number of kW x the number of hours
   0.48 kW x 5 hours = 2.4 kWh

total cost per day = cost per kWh x the number of kWh per day
   20p x 2.4 kWh = 48p

Second calculate the cost of the 12 W bulbs

total amount of kW being used = amount of kW per bulb x 8 bulbs
                             = 0.012 kW x 8 bulbs
                             = 0.096 kW

number of kW hours is the number of kW x the number of hours = 0.096 kW x 5 hours
                                                                    = 0.48 kWh

total cost per day = cost per kWh x the number of kWh per day
$$= 20p \times 0.48 \text{ kWh}$$
$$= 9.6p$$

savings per day = the difference in the two costs
$$= 48p - 9.6p$$
$$= 38.4p$$

8.9) A 2 kW kettle takes 10 minutes to boil. Every day it is boiled 3 times to make cups of coffee.
   a) Calculate the number of kilowatt hours that are required for the kettle to boil every year.
   b) Calculate the cost if each kWh is 20p.

a) The amount of kWh the kettle is used for each day

$$\text{amount of kW the kettle is used for each day} = \frac{2 \text{ kW} \times 10 \text{ minutes} \times 3 \text{ times per day}}{60 \text{ minutes in an hour}}$$
$$= 1 \text{ kWh per day}$$

The amount of kWh the kettle is used for per year is then just 1 kWh per day x 365 days per year

b) amount used each year = 365 kWh per year
total cost = 365 kWh x 20p
$$= 7,300p$$
$$= £73$$

8.10) For the previous question calculate the amount of energy used, in joules, to make cups of coffee per year.

From the previous question we know that the amount of energy used is 365 kWh per year.

energy = power x time

The units are already in this format. What we need are the units in joules and for this we need the time in seconds.

First step: How many joules are in 1 kWh?
energy = time x power

$$1 \text{ kW} = 1,000 \text{ W}; \quad 1W = 1 \text{ Js}^{-1}$$

energy = 60 s min^{-1} x 60 min hr^{-1} x 1,000 Js^{-1}
$$= 3,600,000 \text{ J}$$
$$= 3.6 \text{ MJ per kWh}$$

We use 365 kWh per year. This equates to:
energy per year in joules = 3.6 MJ per kWh x 365 kWh per year
$$= 1.3 \text{ GJ per year}$$

8.11) A power station operates at 45% efficiency, generating an output of 100 MW. Calculate the power available from the fuel that is put into the power station.

total energy in = $\dfrac{\text{useful work done}}{\text{efficiency}}$

= $\dfrac{100 \text{ MW}}{0.45}$

= 222 MW

8.12) A system of power generation has an efficiency of 34%. This system generates 500 MW of energy to transport.
    a) How much energy is lost in the generation process?
    b) How much energy would be generated if the efficiency was to rise to 37%?

a)

total energy in = $\dfrac{\text{useful work done}}{\text{efficiency}}$

= $\dfrac{500 \text{ MW}}{0.34}$

= 1,470 MW

energy lost = total energy in − useful work done
          = 1,470 MW − 500 MW
          = 970 MW

b) useful work done = efficiency x total energy in
                          = 0.37 x 1,470 MW
                          = 544 MW

Bonus Questions 8.1: Charge, Energy, Efficiency and Electrical Power 1

8.1.1) How much current flows in a circuit with 12 Ω of resistance and a supply voltage of 24 V?

8.1.2) 15 amps pass through a resistor of 23 Ω. Calculate the value of the voltage across the resistor.

8.1.3) A circuit is supplied with 3 V. The circuit draws a current of 15 amps. Calculate the value of the resistance of the circuit.

8.1.4) Calculate the charge that passes through an electric circuit component when it is supplied with a current of 3 A for a time period of 60 seconds.

8.1.5) What current must exist for a charge of 300 C to be passed through a light bulb in a time of 5 minutes?

8.1.6) How long does it take for 1,200 coulombs to pass through a point in a circuit that is supplied with 12 A of current?

8.1.7) A voltage of 30 V is applied across a circuit of resistance 2 Ω. How long does it take for a charge of 240 C to move past a point in the circuit?

8.1.8) How much energy is needed to move a charge of 40 C through a voltage of 35 V?

8.1.9) If 500 J of energy was used to move a charge through a voltage of 23 V how much charge has been moved?

8.1.10) Calculate the energy that is needed to move a charge of 3 micro coulombs through a voltage of 120 kV.

8.1.11) It takes 3,400 J of energy to move a charge of 12 C through a voltage. Calculate the value of the voltage through which the charge has been moved.

8.1.12) It would take 76.5 MJ to move 3 coulombs of charge through the largest voltage ever created in a lab. What was the value of the voltage created?

8.1.13) A lightning bolt moves 20 C of charge through a voltage of 120 MV. Calculate the energy needed to perform this.

8.1.14) How much energy is required to power a 100 W light bulb for a period of 2 minutes?

8.1.15) What is the power of an electrical circuit that converts 20 kJ of electrical energy in a time period of 3 minutes?

8.1.16) How much time is needed to convert 1 MJ of electrical energy in a 1.6 kW kettle?

8.1.17) A coal fired power station takes in 38 MJ of energy from coal every second and produces 15 MW of electrical energy. Calculate the efficiency of the power station.

8.1.18) Why is it that any process, such as burning fuel to turn a turbine to make electricity or using a wind turbine to harness energy is never 100% efficient?

8.1.19) A wind turbine produces 500 kW of power. If the wind turbine is 40% efficient then how much energy is actually passing through the wind turbine?

8.1.20) 50 kW of electrical power is supplied to a crane to lift a 10,000 N weight. If the crane operates at 70% efficiency how long will it take to lift a weight of 10,000 N a distance of 20 m?

8.1.21) An electric motor is 90% efficient. It is used to lift a 40 kg mass a height of 50 m in a time of 40 seconds.
   a) Calculate the rate at which work is being done in lifting the mass.
   b) Calculate the power being supplied to the motor.
   c) If the motor has a resistance of 30 Ω what is the value of the current that is travelling through the motor? (use the equation: power = I^2R)

Bonus Questions 8.2: Charge, Energy, Efficiency and Electrical Power 1

8.2.1) Calculate the value of the current flowing in a circuit with 16 Ω of resistance and a supply voltage of 48 V.

8.2.2) What is the value of the voltage across the resistor when 3 amps passes through the resistor of resistance 30 Ω?

8.2.3) Find the resistance of a circuit if, when it is supplied with 10 V, the circuit draws a current of 2 amps.

8.2.4) What amount of charge passes through an electric circuit component when it is supplied with a current of 12 A for a time period of 20 seconds?

8.2.5) If 240 C is to be passed through a light bulb in a time of 4 minutes, find the current that is required to pass through the bulb.

8.2.6) How long does it take for 900 coulombs to pass through a point in a circuit that is supplied with 4 A of current?

8.2.7) 35 V is applied across a resistor of resistance 20 Ω. How long does it take for a charge of 180 C to move past a point in the resistor?

8.2.8) Calculate the amount of energy required to move a charge of 38 C through a voltage of 15 V?

8.2.9) How much charge has been moved if 320 J of energy was used to move the charge through a voltage of 48 V?

8.2.10) Calculate the energy required to move a charge of 12 micro coulombs through a voltage of 60 kV.

8.2.11) If it required 270 J of energy to move an 18 C charge through a voltage, calculate the value of the voltage through which the charge has been moved.

8.2.12) It would take 2.2 GJ to move 18 coulombs of charge through a bolt of lightning. What was the value of the voltage created by the lightning?

8.2.13) I move 3 C of charge through a voltage of 22 MV. What is the energy that this would require?

8.2.14) How much energy is required to power a 2 kW heater for a period of 2 minutes?

8.2.15) Calculate the power of an electrical circuit that converts 200 kJ of electrical energy in a time period of 12 minutes?

8.2.16) How much time is needed to convert 1 MJ of electrical energy with an 800 W hair dryer?

8.2.17) A gas power station takes in 50 MJ of energy from gas every second and produces 22 MW of electrical energy. Calculate the efficiency of the power station.

8.2.18) In lightning, charge travels from a cloud to the ground. If I had a machine on the ground that was able to directly capture lightning as it was created, could it be 100% efficient?

8.2.19) A Solar panel is producing 200 W of electrical energy. If the solar panel is 2 m² and the light intensity landing on the panel is 850 W per square meter, calculate the efficiency of the panel.

8.2.20) A 30 kW lift is designed to carry vehicles to parking locations. If the lift operates at 80% efficiency, how long will it take to lift a car that weighs 14,000 N a distance of 15 m?

8.2.21) A funicular has an electric motor with an efficiency of 85%. It is used to lift a 4,000 kg tram a height of 30 m in a time of 6 minutes.
   a) Calculate the rate at which work is being done in lifting the tram.
   b) Calculate the power being supplied to the motor.
   c) If the motor has a resistance of 30 Ω what is the value of the current that is travelling through the motor? (use the equation power = I^2R)

9. Resistors in Series and Parallel, Voltage Across Resistors

Sapud the Wonderfish

Finding resistance for series and parallel circuits.

SAPUD the Wonderfish

Series: **a**dd them
Parallel: **u**pside **d**own

Resistors in **s**eries are **a**dded:

$$R_{Total} = R_1 + R_2 + R_3 \ldots$$

Resistors in **p**arallel are added **u**pside **d**own:

$$\frac{1}{R_{Total}} = \frac{1}{R_1} + \frac{1}{R_2} + \frac{1}{R_3} \ldots$$

Rusty Robots

This is how we can remember the equation for voltage across a resistor.

Very visible rust over all robots

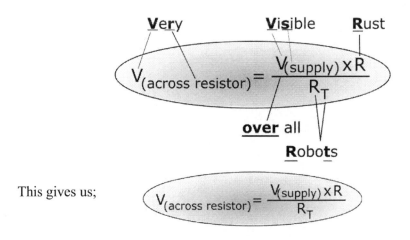

This gives us;

$$V_{(across\ resistor)} = \frac{V_{(supply)} \times R}{R_T}$$

V across resistor – This is the voltage across the resistor you are interested in.
V supply – This is the value, in volts, of the voltage supplied to the circuit.
R – This is the value of the resistance you want to find the voltage across, in ohms.
R_t – This is the value of the total resistance of the circuit.

This equation gives us the voltage across a resistor in a series circuit.

Examples: Resistors in Series and Parallel, Voltage Across Resistors

Example 1) I have a current of 20 amps and a resistance of 3 ohms in a circuit. Calculate the supply voltage.

voltage = current x resistance
       = 20 amps x 3 ohms
       = 60 volts

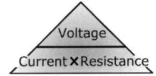

Example 2) I have a resistance of 3 ohms in a circuit supplied by 12 volts. What is the current in the circuit?

current = $\dfrac{\text{voltage}}{\text{resistance}}$

      = $\dfrac{12 \text{ V}}{3 \text{ } \Omega}$

      = 4 amps

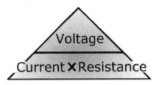

Example 3) What is the resistance of an ammeter and a voltmeter and where are they placed in a circuit?

A voltmeter has an infinite (or very large) resistance. It is placed in parallel with the circuit component.

An ammeter has zero (or very small) resistance. It is placed in series with the circuit component.

Example 4) a) Calculate the total resistance of the following circuit. Each of the resistors has a resistance of 3 ohms.

The resistors here are arranged in series.

R_t = 3 ohms + 3 ohms + 3 ohms
    = 9 ohms

b) If the supply voltage is 12 volts calculate the value of the current.

current = $\dfrac{\text{voltage}}{\text{resistance}}$

      = $\dfrac{12 \text{ V}}{9 \text{ } \Omega}$

      = 1.33 amps

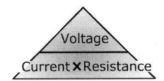

Example 5) In the following example of a parallel circuit each of the resistors have a resistance of 3 ohms.
a) Calculate the total resistance of the circuit.

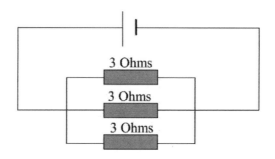

The resistors here are arranged in parallel.

$$\frac{1}{R_t} = \frac{1}{3} + \frac{1}{3} + \frac{1}{3}$$

$$= \frac{1}{1}$$

$$\boxed{\frac{1}{R_{Total}} = \frac{1}{R_1} + \frac{1}{R_2} + \frac{1}{R_3} \ldots}$$

$R_t = 1$ ohm

b) If there is a supply voltage of 12 volts calculate the current in the circuit.

current = $\frac{\text{voltage}}{\text{resistance}}$

= $\frac{12 \text{ V}}{1 \text{ }\Omega}$

= 12 amps

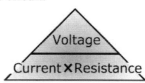

Example 6) I have three 2 ohm resistors. Calculate the total resistance if I place them a) in series, b) in parallel

a)

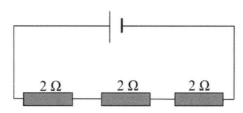

$R_t = 2$ ohms + 2 ohms + 2 ohms
 = 6 ohms

$$\boxed{R_{Total} = R_1 + R_2 + R_3 \ldots}$$

b)

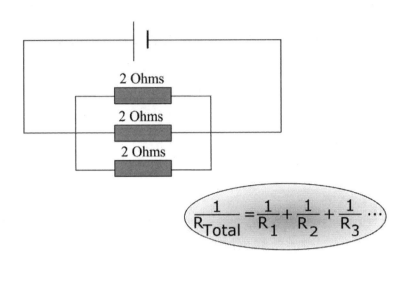

$$\frac{1}{R_t} = \frac{1}{2} + \frac{1}{2} + \frac{1}{2}$$
$$= \frac{3}{2}$$
$$\frac{R_t}{1} = \frac{2}{3}$$
$$\text{resistance} = \frac{2}{3}$$
$$= 0.667 \, \Omega$$

$$\frac{1}{R_{Total}} = \frac{1}{R_1} + \frac{1}{R_2} + \frac{1}{R_3} \ldots$$

Example 7) There are four 3 ohm resistors in series in a circuit. The circuit is supplied by a 12 volt power supply. Calculate the value of the voltage across each resistor.

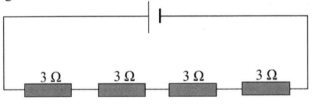

Step 1: Calculate the total resistance in the circuit.

$R_t = 3$ ohms $+ 3$ ohms $+ 3$ ohms $+ 3$ ohms
    $= 12$ ohms

$$R_{Total} = R_1 + R_2 + R_3 \ldots$$

Step 2: Use the equation that tells you the voltage across the resistor.

resistor 1 = 3 ohms

voltage across resistor 1 $= \dfrac{12 \text{ V} \times 3 \, \Omega}{12 \, \Omega}$

$$V_{(across\ resistor)} = \frac{V_{(supply)} \times R}{R_T}$$

$= 3$ volts

As resistors 2, 3 and 4 are also 3 ohms they also each have 3 volts across them. The voltage goes where it is needed in the circuit according to the rules that you have learned. We can also see that all of the voltage is used moving the charge through the circuit. There are 4 resistors each with 3 volts across them with a total of 4 x 3 V = 12 V. This is the supply voltage of the circuit!

Using the equation will give the same answer.
The voltage across each resistor is 3 volts.

Example 8) There are four 2 ohm resistors in series in a circuit. The circuit is supplied by a 12 volt power supply. Calculate the voltage across the individual resistors.

Step 1: Calculate the total resistance in the circuit.

R_t = 2 ohms + 2 ohms + 2 ohms + 2 ohms
   = 8 ohms

$$R_{Total} = R_1 + R_2 + R_3 ...$$

Step 2: Use the equation that tells you the voltage across the resistor.

$V_{across\ resistor} = V_{supply} (R / R_t)$

resistor 1 = 2 ohms

voltage across resistor 1 = $\dfrac{12\ V \times 2\ \Omega}{8\ \Omega}$

= 3 volts.

$$V_{(across\ resistor)} = \dfrac{V_{(supply)} \times R}{R_T}$$

Resistor 2, 3 and 4 are also 2 ohms.

Using the equation will give the same answer.

The voltage across each resistor is 3 volts.

Example 9) I have 4 resistors in series of resistances 1 ohm, 2 ohms, 3 ohms and 4 ohms. There is a supply voltage of 20 volts. Calculate the voltage across each resistor.

Step 1: Calculate the total resistance in the circuit.

R_t = 1 ohm + 2 ohms + 3 ohms + 4 ohms
   = 10 ohms

$$R_{Total} = R_1 + R_2 + R_3 ...$$

Step 2: Use the equation that tells you the voltage across the resistor.

resistor 1 = 1 ohm

voltage across resistor 1 = $\dfrac{20\ V \times 1\ \Omega}{10\ \Omega}$

= 2 volts

$$V_{(across\ resistor)} = \dfrac{V_{(supply)} \times R}{R_T}$$

resistor 2 = 2 ohms

voltage across resistor = $\dfrac{20\ V \times 2\ \Omega}{10\ \Omega}$

= 4 volts

resistor 3 = 3 ohms

voltage across resistor = $\dfrac{20\text{ V} \times 3\,\Omega}{10\,\Omega}$

= 6 volts

resistor 4 = 4 ohms

voltage across resistor = $\dfrac{20\text{ V} \times 4\,\Omega}{10\,\Omega}$

= 8 volts

Example 10) In the following circuit each of the resistors has a resistance of 2 ohms. The supply voltage is 12 volts. Calculate the voltage across each of the resistors.

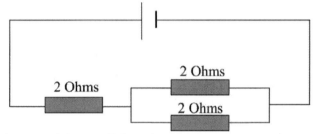

Step 1: Calculate the total resistance of the parallel section.

$\dfrac{1}{R_t} = \dfrac{1}{2} + \dfrac{1}{2}$

$= \dfrac{1}{1}$

$= \dfrac{1}{1}$

$= 1\,\Omega$

$\dfrac{1}{R_{Total}} = \dfrac{1}{R_1} + \dfrac{1}{R_2} + \dfrac{1}{R_3} \cdots$

This means that the following 2 sections of circuit have the same resistance.

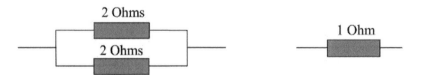

This is the value of a resistor that we could put in place of the 2 resistors in parallel.

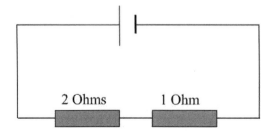

We are now in the position to calculate the total resistance of the circuit.

R_t = 2 ohms + 1 ohm
  = 3 ohms

$$R_{Total} = R_1 + R_2 + R_3 ...$$

Step 2: We can now calculate the voltage across the resistors.

voltage across 2 Ω resistor = $\dfrac{12 \text{ V} \times 2 \text{ Ω}}{3 \text{ Ω}}$

$$V_{(across\ resistor)} = \dfrac{V_{(supply)} \times R}{R_T}$$

  = 8 V

V across the parallel resistors is calculated because they can be exchanged for a single 1 ohm resistor.

voltage across parallel resistors = $\dfrac{12 \text{ V} \times 1 \text{ Ω}}{3 \text{ Ω}}$

  = 4 volts

Both the parallel resistors will have the 4 volts across them.

Questions: Resistors in Series and Parallel, Voltage Across Resistors

9.1) I have three 2 ohm resistors in series in a circuit. Calculate the total resistance of the circuit.

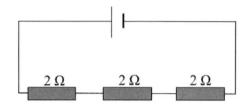

9.2) I have three 4 ohm resistors in parallel in a circuit. Calculate the total resistance that this creates.

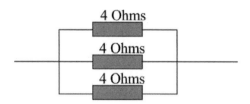

9.3) I build a circuit to measure the current through a metal wire. Once I have done this I begin to heat the metal wire. As the temperature increases the current in the wire gets smaller. What is happening in the wire?

9.4) Calculate the current produced when a 12 volt power supply is used across a circuit with a resistance of 3 ohms.

9.5) I have three 1 ohm resistors. Calculate the total resistance if the resistors are arranged in series and then if they are arranged in parallel.

9.6) I have three resistors 1 ohm, 100 ohms and 1,000 ohms. They are arranged in parallel. Calculate the final resistance and then comment on the result.

9.7) I have 2 identical resistors arranged in parallel. When I have a 10 volt power supply across them I get a current of 5 amps in the circuit. Calculate the value of the individual resistors.

9.8) There are 3 resistors in parallel. Each has a value of 3 ohms. After the parallel section of the circuit they are connected to a 3 ohm resistor in series. What is the value of the final resistance?

9.9) There are seven 2 ohm resistors in parallel in a circuit. The circuit is supplied by a 10 volt DC power supply. What is the voltage across each of the resistors?

9.10) I have four 2 ohm resistors in parallel and a 2 ohm resistor connected in series after that. The circuit is connected to a 12 volt power supply. What is the voltage across each of the resistors?

Answers: Resistors in Series and Parallel, Voltage Across Resistors

9.1) I have three 2 ohm resistors in series in a circuit. Calculate the total resistance of the circuit.

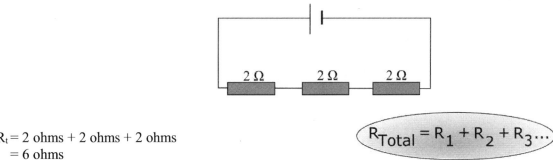

R_t = 2 ohms + 2 ohms + 2 ohms
= 6 ohms

$$R_{Total} = R_1 + R_2 + R_3 ...$$

9.2) I have three 4 ohm resistors in parallel in a circuit. Calculate the total resistance that this creates.

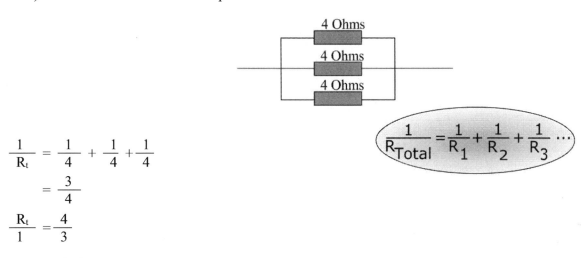

$$\frac{1}{R_t} = \frac{1}{4} + \frac{1}{4} + \frac{1}{4}$$

$$= \frac{3}{4}$$

$$\frac{R_t}{1} = \frac{4}{3}$$

$$\frac{1}{R_{Total}} = \frac{1}{R_1} + \frac{1}{R_2} + \frac{1}{R_3} ...$$

R_t = 1.33 ohms

9.3) I build a circuit to measure the current through a metal wire. Once I have done this I begin to heat the metal wire. As the temperature increases the current in the wire gets smaller. What is happening in the wire?

$$\text{current} = \frac{\text{voltage}}{\text{resistance}}$$

Voltage / Current × Resistance

Voltage is constant. The only thing that is changing the value of the current is the resistance. If the current is decreasing the resistance must be increasing.

(Interestingly, the resistance in a normal metallic conductor will increase as the temperature increases. This is because all the atoms vibrate about their fixed points more as their kinetic energy increases, because kinetic energy is proportional to temperature. So as the electrons try to move past, it gets more difficult and takes more energy – like moving through a crowded room.)

9.4) Calculate the current produced when a 12 volt power supply is used across a circuit with a resistance of 3 ohms.

current = $\dfrac{\text{voltage}}{\text{resistance}}$

current = $\dfrac{12\ V}{3\ \Omega}$

current = 4 amps

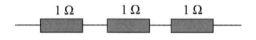

9.5) I have three 1 ohm resistors. Calculate the total resistance if the resistors are arranged in series and then if they are arranged in parallel.

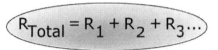

R_t = 1 ohm + 1 ohm + 1 ohm
= 3 ohms
total resistance for the resistors in series = 3 ohms

$R_{Total} = R_1 + R_2 + R_3 \ldots$

The total resistance for the resistors in parallel can be calculated as follows:

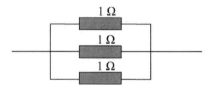

$\dfrac{1}{R_t} = \dfrac{1}{1} + \dfrac{1}{1} + \dfrac{1}{1}$

$= \dfrac{3}{1}$

$\dfrac{R_t}{1} = \dfrac{1}{3}$

$\dfrac{1}{R_{Total}} = \dfrac{1}{R_1} + \dfrac{1}{R_2} + \dfrac{1}{R_3} \ldots$

R_t = 0.33 ohms
total resistance for the parallel resistors = 0.33 ohms

9.6) I have three resistors 1 ohm, 100 ohms and 1,000 ohms. They are arranged in parallel. Calculate the final resistance and then comment on the result.

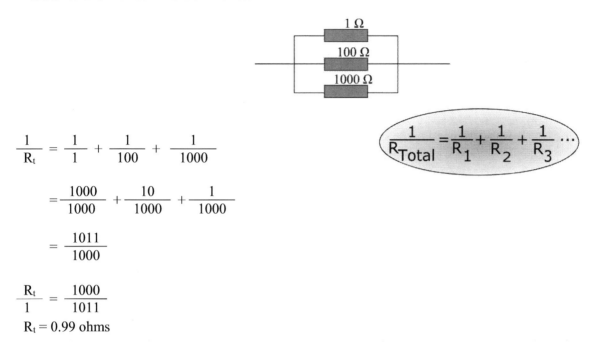

$$\frac{1}{R_t} = \frac{1}{1} + \frac{1}{100} + \frac{1}{1000}$$

$$= \frac{1000}{1000} + \frac{10}{1000} + \frac{1}{1000}$$

$$= \frac{1011}{1000}$$

$$\frac{R_t}{1} = \frac{1000}{1011}$$

$$R_t = 0.99 \text{ ohms}$$

The total resistance of resistors arranged in parallel is always smaller than the smallest of the resistors.

9.7) I have 2 identical resistors arranged in parallel. When I have a 10 volt power supply across them I get a current of 5 amps in the circuit. Calculate the value of the individual resistors.

$$\text{resistance} = \frac{\text{voltage}}{\text{current}}$$

$$= \frac{10 \text{ V}}{5 \text{ A}}$$

$$= 2 \text{ ohms}$$

Working backwards from the resistors in parallel:

We know the final value for the total resistance = 2 ohms and we know that both of the resistors are of the same value.

$$\frac{1}{2} = \frac{1}{R} + \frac{1}{R}$$

$$= \frac{2}{R}$$

$$\frac{2}{1} = \frac{R}{2}$$

$$4 = R$$

$$R = 4 \text{ ohms}$$

9.8) There are 3 resistors in parallel. Each has a value of 3 ohms. After the parallel section of the circuit they are connected to a 3 ohm resistor in series. What is the value of the final resistance?

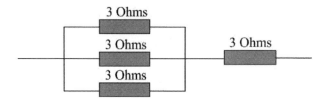

Step 1: Calculate the resistance of the parallel section.

$$\frac{1}{R_t} = \frac{1}{3} + \frac{1}{3} + \frac{1}{3}$$

$$= \frac{3}{3}$$

$$\frac{R_t}{1} = \frac{3}{3}$$

$$\frac{1}{R_{Total}} = \frac{1}{R_1} + \frac{1}{R_2} + \frac{1}{R_3}\ldots$$

$R_t = 1$ ohm

Step 2: Calculate the resistance of the total circuit. The value for the parallel section is the same value as a single resistor that could replace all three of the parallel resistors.

We are now in the position to calculate the total resistance of the circuit.

$R_t = 1$ ohm $+ 3$ ohms
$R_t = 4$ ohms

$$R_{Total} = R_1 + R_2 + R_3\ldots$$

The total resistance of the circuit is 4 ohms

9.9) There are seven 2 ohm resistors in parallel in a circuit. The circuit is supplied by a 10 volt DC power supply. What is the voltage across each of the resistors?

The 10 V supply is connected across all of the resistors. The voltage across each of the resistors is the same. It is 10 V.

9.10) I have four 2 ohm resistors in parallel and a 2 ohm resistor connected in series after that. The circuit is connected to a 12 volt power supply. What is the voltage across each of the resistors?

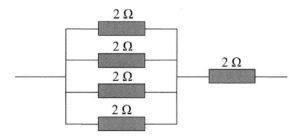

Step 1: Calculate the resistance of the parallel section.

$$\frac{1}{R_t} = \frac{1}{2} + \frac{1}{2} + \frac{1}{2} + \frac{1}{2}$$

$$= \frac{4}{2}$$

$$\frac{R_t}{1} = \frac{2}{4}$$

$$\frac{1}{R_{Total}} = \frac{1}{R_1} + \frac{1}{R_2} + \frac{1}{R_3} \cdots$$

$R_t = 0.5$ ohms

Step 2: Calculate the resistance of the total circuit. The value for the parallel section is the same value as a single resistor that could replace all three of the parallel resistors.

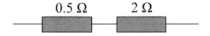

We are now in the position to calculate the total resistance of the circuit.

R_t = 0.5 ohms + 2 ohms
    = 2.5 ohms

$$R_{Total} = R_1 + R_2 + R_3 ...$$

The total resistance of the circuit is 2.5 ohms.

We are now in a position to calculate the voltage across the resistors.

voltage across parallel resistors = $\dfrac{12 \text{ V} \times 0.5 \, \Omega}{2.5 \, \Omega}$

= 2.4 volts

$$V_{(across\ resistor)} = \dfrac{V_{(supply)} \times R}{R_T}$$

voltage across series resistor = $\dfrac{12 \text{ V} \times 2 \, \Omega}{2.5 \, \Omega}$

= 9.6 volts

Bonus Questions 9.1: Resistors in Series and Parallel, Voltage Across Resistors

9.1.1) Calculate the resistance of a 10, 20, 30 and a 50 Ω resistor all connected in series.

9.1.2) I have 6 identical resistors. If the combined resistance is 180 Ω when the resistors are connected together in series, calculate the resistance of a single resistor.

9.1.3) Calculate the resistance if the resistors from the previous question had all been connected in parallel.

9.1.4) How can I connect 4 resistors each with a value of 2 Ω so that my final value for the combined resistance is also 2 Ω? Why might this be useful?

9.1.5) A wire has a resistance of 500 Ω per metre. What length of wire would I need in order to create a resistance of 940 Ω?

9.1.6) Using the wire from the previous question what would the resistance of the following design be?
Two sections of wire, each of length 3 m connected in parallel.

9.1.7) A 2 Ω and an 8 Ω resistor are connected in parallel. This combination is then connected in series with a string of resistors of value 3, 4 and 5 Ω which are all connected in series.
 a) Calculate the total resistance of the circuit.
 b) A 100 V supply is then connected across the ends of the circuit. Calculate the value of the current that would be provided to the circuit.

9.1.8) I have a 2.5 Ω resistor that is broken. I need to create an arrangement of resistors to replace the 2.5 Ω resistor. I have several 1 Ω resistances. How could I arrange them to get 2.5 Ω?

9.1.9) I have 2 light bulbs each marked 50 W but one of them is clearly much brighter.
 a) Which one is more efficient?
 b) Where does the extra energy that is converted go to in the other bulb?

9.1.10) Four 4 Ω resistors are connected in parallel to make a circuit. A voltage is connected across this circuit. Once the supply voltage is connected it produces a current of 17 amps. Calculate the value of the voltage that has been connected.

9.1.11) I attach 3 identical resistors in parallel. The final resistance of the arrangement is 3 Ω. What is the value of an individual resistor?

9.1.12) What are the resistances of the following?
 a) Two 2 Ω resistors connected in parallel.
 b) Three 3 Ω resistors connected in parallel.
 c) Four 4 Ω resistors connected in parallel.

9.1.13) Three 4 Ω resistors are connected in parallel. This circuit is then attached to a switch of resistance 1 Ω. Calculate the final resistance of the circuit.

9.1.14) Two 12 Ω resistors are connected in series. A 12 volt power supply is then connected to this circuit. Calculate the value of the voltage across each resistor.

9.1.15) A 5 Ω and a 9 Ω resistor are connected in series. 90 volts is then connected across the circuit. Calculate the value of the voltage across the 5 Ω resistor.

9.1.16) A 6 Ω and an 8 Ω resistance are connected in parallel. This circuit is then connected to a 5 Ω resistance in series. A 20 V power supply is then connected across the complete circuit. What is the value of the potential difference across the 5 Ω resistor?

9.1.17) A 4 Ω resistor is connected in parallel with a 2 Ω resistor. These are then connected in series with a 1 Ω resistor and the circuit is attached to a 12 V power supply.
   a) Calculate the value of the voltage across the 4 Ω resistor.
   b) Calculate the value of the voltage across the 1 Ω resistance.
   c) What is the value of the voltage across the 2 Ω resistor?

9.1.18) A 1 Ω, 4 Ω and 5 Ω resistor are all connected in series and then a 20 V power supply is connected across the circuit. Calculate the following.
   a) The voltage across the 1 Ω resistor.
   b) The voltage across the 4 Ω resistor.
   c) The voltage across the 5 Ω resistor.
   d) If you add up all of these voltages do you get the supply voltage?
   e) Individually, using the values that you have calculated above, calculate the value of the current in each of the resistors.

9.1.19) A 1 Ω, 4 Ω and 5 Ω resistor are all connected in parallel. A 20 V power supply is connected across the circuit. Calculate the following.
   a) The voltage across the 1 Ω resistor.
   b) The voltage across the 2 Ω resistor.
   c) The voltage across the 5 Ω resistor.
   d) What do you notice about the voltages?

9.1.20) Which of the following happens when a lighting circuit is complete and the electricity is being supplied to a light bulb? Why does the light become bright?
   a) The light being emitted shows that electrons are being used up.
   b) Energy is used up and that is why it is bright.
   c) Electrical energy makes the bulb bright.
   d) The energy is changing from electrical to light and heat in the bulb.

Bonus Questions 9.2: Resistors in Series and Parallel, Voltage Across Resistors

9.2.1) What is the total resistance when a 20, 30, 30 and a 60 Ω resistor are connected in series?

9.2.2) There are 3 identical resistors all connected in series. If the combined resistance is 108 Ω, what is the resistance of a single resistor?

9.2.3) What would the value of the total resistance be if the resistors from the previous question had all been connected in parallel?

9.2.4) I have 4 identical resistors. I create two sets of two resistors in series and then put the two pairs in parallel. How does the total resistance compare to the resistance of each individual resistor?

9.2.5) I need to create a resistance of 630 Ω using wire that has a resistance of 380 Ω per meter. What length of wire would I need?

9.2.6) Using the wire from the previous question what would the resistance of the following design be?
Two sections of wire, each 2 metres long, connected in parallel.

9.2.7) A string of resistors of values 1, 2 and 4 Ω are all connected in series. There is also a 4 Ω and a 2 Ω resistor connected in parallel. This combination is then connected in series with the string of resistors.
   a) Calculate the total resistance of the circuit.
   b) A 40 V supply is then connected across the ends of the circuit. Calculate the value of the current that would be provided to the circuit.

9.2.8) I have several 2 Ω resistors. How could I arrange them to get a 3 Ω resistance?

9.2.9) I measure the light output from 2 televisions. Both of them have the same power requirements but the television on the left is much brighter. Both of the televisions have had their sound muted.
   a) Which one is more efficient?
   b) Where does the extra energy go to in the less efficient television?

9.2.10) Five 5 Ω resistors are connected in parallel to make a circuit and a voltage is connected across this circuit. Once the voltage has been connected across the ends of the circuit a current of 5 A is measured. Calculate the value of the voltage that has been connected.

9.2.11) 4 identical resistors are connected in parallel. The final resistance of the arrangement is 5 Ω. What is the value of an individual resistor?

9.2.12) Calculate the resistances of the following circuits.
   a) Three 2 Ω resistors connected in parallel.
   b) Two 2 Ω resistors in parallel then connected with another 2 Ω resistance in series.
   c) Three 2 Ω resistors in series.

9.2.13) Four 1 Ω resistors are connected in parallel. This circuit is then attached to a switch of resistance 2 Ω. Calculate the final resistance of the circuit and switch.

9.2.14) Two 6 Ω resistors are connected in series. A 24 volt power supply is then connected to this circuit. Calculate the value of the voltage across each resistor.

9.2.15) A 4 Ω and a 10 Ω resistor are connected in series. 80 volts is then connected across the circuit. Calculate the value of the voltage across the 10 Ω resistor.

9.2.16) A 10 Ω and a 12 Ω resistance are connected in parallel. This circuit is then connected to a 4 Ω resistance in series. A 24 V power supply is then connected across the complete circuit. What is the value of the potential difference across the 4 Ω resistor?

9.2.17) A 12 Ω resistor is connected in parallel with a 6 Ω resistor. These are then connected in series with a 1 Ω resistor and the circuit is attached to a 12 V power supply.
   a) Calculate the value of the voltage across the 6 Ω resistor.
   b) Calculate the value of the voltage across the 1 Ω resistance.
   c) What is the value of the voltage across the 12 Ω resistor?

9.2.18) A 3 Ω, 4 Ω and 6 Ω resistor are all connected in series and then a 12 V power supply is connected across the circuit. Calculate the following.
   a) The voltage across the 3 Ω resistor.
   b) The voltage across the 4 Ω resistor.
   c) The voltage across the 6 Ω resistor.
   d) If you add up all of these voltages do you get the supply voltage?
   e) Individually, using the values that you have calculated above, calculate the value of the current in each of the resistors.

9.2.19) A 3 Ω, 4 Ω and 6 Ω resistor are all connected in parallel. A 12 V power supply is connected across the circuit. Calculate the following.
   a) The voltage across the 3 Ω resistor.
   b) The voltage across the 4 Ω resistor.
   c) The voltage across the 6 Ω resistor.
   d) What do you notice about the voltages?

9.2.20) Which of the following happens when a heating circuit is complete and the electricity is being supplied to an electric heater? Why does the heater become warm?
   a) Electrical energy is changed to heat energy as the electrons are used up.
   b) Electrical energy is changed to heat energy but the electrons are not used up.
   c) Electrical energy is given out as the electrons are used up.
   d) Electrical energy is given out but the electrons are not used up.

10. Electrical Power 2 and Transformers

The Power of Imagination

It's **r**eally **i**nteresting
in Venice.
(In Venus?)
Venus??? **Too over r**un
(with alien mice.)

Equations for electrical power (the **power** of imagination)

It's really interesting
in Venice.
(In Venus?)
Venus??? Too over run
(with alien mice.)

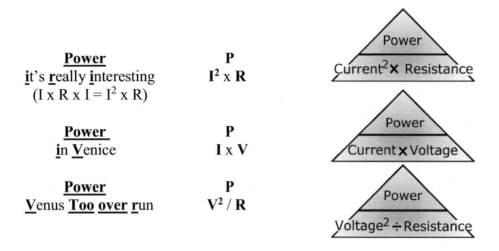

Power **i**t's **r**eally **i**nteresting (I x R x I = I² x R)	P	I² x R
Power **i**n **V**enice	P	I x V
Power **V**enus **T**oo **o**ver **r**un	P	V² / R

Just a quick note. In school you are probably learning all of these as the three equations of power.

These are:
- Power = I²R → this is a very important equation when we deal with electricity being transported across wires and we wish to calculate energy lost due to resistance.
- Power = V²/R
- Power = IV.

The first and third equations are in the same form and as you can see with the second equation covering over power immediately gives you V²/R. So this equation is exactly the same as the one learned in school.

Equations for Transformers: Vans and Vampires

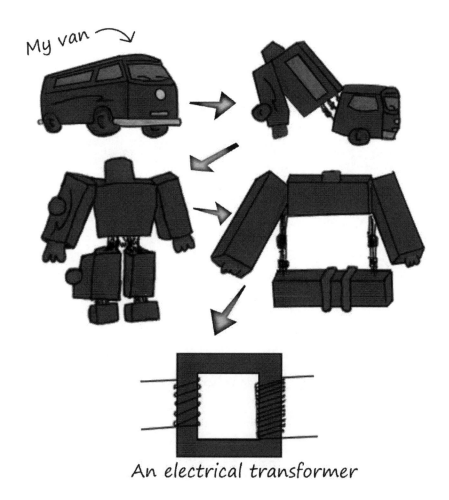

An electrical transformer

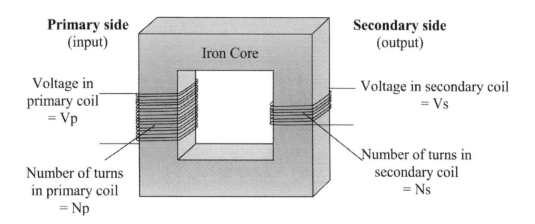

I prefer my **first van** over my **second van.**

From this we can get two important equations for transformers. First we arrange the above as corners in a square. (Remember that another word for first is primary.)

I prefer my **first van** over my **second van.**

To find our 2 equations that we need we can either join these from left to right. You will see that this separates the left and right side with a large = sign. Or we can join them diagonally, giving us a multiplying sign.

Let's examine the square on the left first, the one with the large = sign.

Following the lines that join each other and setting both sides equal to each other gives us one of our useful equations for transformers.

$$\frac{V_p}{V_s} = \frac{N_p}{N_s}$$

While using the square on the right, the one with the large multiplication sign, gives us our second important equation:

$$N_p \times V_s = N_s \times V_p \qquad \frac{V_p \times N_s}{V_s \times N_p}$$

Both of these equations are useful!

But what if you remember these equations in a different order?

As long as the small letters on the top row match (are the same) and the large letters on the top and bottom of each side of the equation match then it will work!

So all of the following will give you a valid set of equations that you can use to answer the questions and it wouldn't matter which one you used. The same rules with the equal sign and the multiplying sign both work.

 or 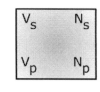 or

So how do you remember this? Just remember that the lower case letters have to go on the same line. The capital letters that are on the left side should be the same (either both V's on the left or both N's on the left) and the ones on the right side should be the remaining letter.

If you want another way of remembering this equation…

Nosey **V**ampires **vi**sit **Na**ples

$$\frac{N_s \times V_p}{V_s \times N_p}$$

'You don't get anything for free!'

This rule of energy conservation is one that people have been claiming for years to have broken, creating perpetual motion machines that allow us to get energy for free with someone sending you the plans for a small fee. The internet is full of such charlatanism and it is always wise to be wary of these things.

In the case of the transformers, we (at this level) assume that they are 100% efficient and all of the energy that we put into them comes out the other side.

This is then: power in = power out $\boxed{Power_{in} = Power_{out}}$

And because: power = current x voltage

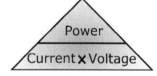

You get: current in x voltage in = current out x voltage out

$$\frac{I_{in} \times V_{in}}{I_{out} \times V_{out}}$$

$$\boxed{\frac{Current_{in} \times Voltage_{in}}{Current_{out} \times Voltage_{out}}}$$

Examples: Electrical Power 2 and Transformers

Example 1) I have a light bulb of resistance 4 ohms and supply voltage 30 volts. Calculate the energy converted in the light bulb:
  a) per second
  b) per minute

a) power = $\dfrac{\text{voltage} \times \text{voltage}}{\text{resistance}}$

= $\dfrac{30\text{ V} \times 30\text{ V}}{4\,\Omega}$

= 225 watts

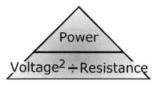

energy = time x power
= 1 s x 225 watts
= 225 joules

b) power = 225 watts

energy = time x power
= 60 s x 225 W
= 13,500 joules
= 13.5 kJ

Example 2) I supply a power of 1,000 watts to a transformer. The voltage is stepped up from its previous value to 100 volts. It is then transferred across a field and then the voltage is stepped back down again.
  a) Calculate the current in the cables. The cables have a resistance of 4 ohms.
  b) Calculate the power converted to heat in the cables (this is often referred to as the power lost).

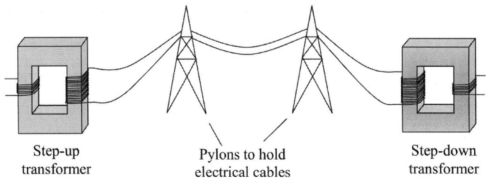

Step-up transformer   Pylons to hold electrical cables   Step-down transformer

a) Find the value of the current in the wires.

current = $\dfrac{\text{power}}{\text{voltage}}$

= $\dfrac{1{,}000\text{ W}}{100\text{ V}}$

= 10 amps

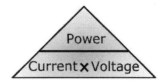

b) Find the value of the power converted to heat in the cables.

power = current x current x resistance
= 10 amps x 10 amps x 4 ohms
= 400 watts

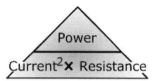

Example 3) Repeat the question with a voltage of 1,000 volts and then repeat it again with a voltage of 10,000 volts.

Repeating with voltage of 1,000 V gives:

current = $\dfrac{\text{power}}{\text{voltage}}$

= $\dfrac{1{,}000 \text{ W}}{1{,}000 \text{ V}}$

= 1 amp

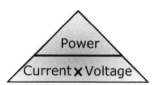

power = current x current x resistance
= 1 amp x 1 amp x 4 ohms
power lost as heat = 4 watts

Repeating with a voltage of 10,000 V gives:

current = $\dfrac{\text{power}}{\text{voltage}}$

= $\dfrac{1{,}000 \text{ W}}{10{,}000 \text{ V}}$

= 0.1 amps

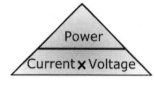

power = current x current x resistance
= 0.1 amp x 0.1 amp x 4 ohms
power lost as heat = 0.04 watts

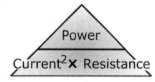

Why do you think that electricity is transported at high voltage?

As the voltage increases the current decreases (because the power supplied is a constant value). Power lost = $I^2 R$ so as the current decreases the power lost as heat decreases proportional to the square of the current. This means that if the current decreases by ½ the power lost will decrease by ¼. If the current decreases to 1/10 of its original value, then the power lost as heat will decrease to 1/100 of its original value.

(Summary: high voltage = low current = low power loss)

Example 4) I have a transformer with 10 kW of power supplied to the primary coil. How much power is supplied from the secondary coil?

power in = power out
= 10 kW
power out = 10 kW

$$\text{Power}_{in} = \text{Power}_{out}$$

Example 5) I have a transformer with a supply voltage of 100 volts. It has 30 turns on the primary coil and 300 turns on the secondary coil. Calculate the output voltage.

$$V_s = \frac{N_s \times V_p}{N_p}$$

$$= \frac{300 \text{ turns} \times 100 \text{ V}}{30 \text{ turns}}$$

$$= 1{,}000 \text{ volts}$$

$$\frac{V_p \times N_s}{V_s \times N_p}$$

Example 6) A transformer has 300 coils of wire on the primary coil. It takes a voltage of 230 V and changes it to a voltage of 3,000 volts at the secondary coil. Calculate the number of turns on the secondary side of the transformer.

$$N_s = \frac{N_p \times V_s}{V_p}$$

$$= \frac{300 \text{ turns} \times 3{,}000 \text{ V}}{230 \text{ volts}}$$

$$= 3{,}913 \text{ turns}$$

$$\frac{V_p \times N_s}{V_s \times N_p}$$

Example 7) I have a supply voltage of 230 volts and I need a voltage of 12 volts to run my computer. Suggest a suitable number of turns on the transformer.

$$\frac{230 \text{ V}}{12 \text{ V}} = \frac{N_p}{N_s}$$

N_p = 230 turns

$$\frac{V_p}{V_s} = \frac{N_p}{N_s}$$

Number of turns on the primary coil = 230 turns

N_s = 12 turns

Number of turns on the secondary coil = 12 turns

Although as long as the ratio is the same the answer is correct i.e. N_p = 115 and N_s = 6 or N_p = 460 and N_s = 24.

Example 8) There is a supply voltage of 300 volts at the primary coil of my transformer. The output coil supplies a voltage of 30 volts. If the input current is 10 amps calculate the value of the output current.

power in = power out

$I_{in} \times V_{in} = I_{out} \times V_{out}$

$I_{out} = \dfrac{I_{in} \times V_{in}}{V_{out}}$

$= \dfrac{10 \text{ A} \times 300 \text{ V}}{30 \text{ V}}$

$\boxed{\text{Power}_{in} = \text{Power}_{out}}$

$\boxed{\dfrac{\text{Current}_{in} \times \text{Voltage}_{in}}{\text{Current}_{out} \times \text{Voltage}_{out}}}$

current out = 100 amps

Example 9) A step up transformer changes a voltage of 20 volts to a voltage of 4,000 volts.

a) Suggest a possible number of turns for the primary and secondary coils of the transformer.

$\dfrac{20 \text{ V}}{4{,}000 \text{ V}} = \dfrac{N_p}{N_s}$

$N_p = 20$

$\boxed{\dfrac{V_p}{V_s} = \dfrac{N_p}{N_s}}$

The number of turns on the primary coil = 20 turns

$N_s = 4{,}000$

The number of turns on the secondary coil = 4,000 turns

b) The power supplied to the transformer is 3,000 watts. Assuming that the transformer is 100% efficient calculate the input and output currents.

current = $\dfrac{\text{power}}{\text{voltage}}$

current in = $\dfrac{3{,}000 \text{ W}}{20 \text{ V}}$

= 150 amps

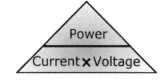

power in = power out

current out = $\dfrac{3{,}000 \text{ W}}{4{,}000 \text{ V}}$

= 0.75 amps

$\boxed{\text{Power}_{in} = \text{Power}_{out}}$

Example 10) The transformer receives a current of 3 amps and a voltage of 230 volts on the input coil. The windings of the transformer are 3,000 on the primary coil and 2,000 on the secondary coil.
   a) Calculate the output voltage and the output power.
   b) Calculate the output current.
   c) If the transformer is only 98 percent efficient. Where does the missing 2% power go?

a)
$$V_s = \frac{N_s \times V_p}{N_p}$$

$$= \frac{2{,}000 \text{ turns} \times 230 \text{ V}}{3{,}000 \text{ turns}}$$

$$= 153 \text{ volts}$$

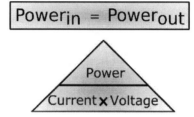

power in = power out

input power = current x voltage
            = 3 amps x 230 volts
            = 690 watts

output power = 690 watts

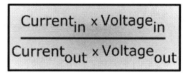

b) $I_{out} = \dfrac{I_{in} \times V_{in}}{V_{out}}$

$= \dfrac{3 \text{ A} \times 230 \text{ V}}{153 \text{ V}}$

current out = 4.51 amps

c) The missing power is lost as heat.

Questions: Electrical Power 2 and Transformers

10.1) Calculate the total amount of energy converted in a 2 kW kettle:
   a) per second
   b) in the 5 minutes it takes to boil

10.2) An electrical appliance converts 360,000 joules of electrical energy every hour when it is switched on. Calculate its power rating.

10.3) How much energy does a 60 W light bulb that is turned on for 3 hours a day use in one year?

10.4) What is the difference between a step up transformer and a step down transformer? Use this information to comment on the number of windings on a transformer.

10.5) I need to design a transformer that will allow me to get 12 volts from a 230 volt power supply. Suggest an appropriate number of turns.

10.6) A transformer has 2 turns on its primary coil and 2,000 turns on its secondary coil. The input voltage is 230 volts. Calculate the output voltage.

10.7) A transformer has an output voltage of 10,000 volts. It has an input voltage of 100,000 volts. The number of turns on the primary coil is 5,000. How many turns are there on the secondary coil?

10.8) A voltage of 10,000 volts is changed down to a voltage of 5,000 volts using a transformer. The power supplied to the transformer at the primary coil is 1,000 watts. The electricity is then transmitted along a cable of resistance 2 ohms.
   a) Calculate the power that is lost in the cable.
   b) Calculate the power that would have been lost in the cable if the transformer was connected the other way around.
   c) Use this answer to comment on why electricity is transported across the country at high voltage.

10.9) There is a supply voltage of 230 volts at the primary coil of my transformer. The output coil supplies a voltage of 230,000 volts. If the input current is 10 amps, calculate the value of the output current.

10.10) 2 transformers are connected in series. A step down transformer is followed by a step up transformer. The step down transformer has 2,000 windings on the primary coil and 250 windings on the secondary coil. The step up transformer has 400 windings on the primary coil and 40,000 on the secondary coil. A supply voltage of 40 volts is fed into the primary coil of the step down transformer. It is then fed directly from the output of the step down transformer to the input of the step up transformer.

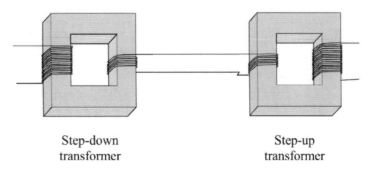

Step-down transformer   Step-up transformer

Calculate the final output voltage from the step up transformer.

Answers: Electrical Power 2 and Transformers

10.1) Calculate the total amount of energy converted in a 2 kW kettle:
   a) per second
   b) in the 5 minutes it takes to boil

a) energy = time x power
         = 1 s x 2,000 W
         = 2,000 J

b) energy = time x power
         = 5 min x 60 s per min x 2,000 W
         = 600,000 J
         = 0.6 MJ

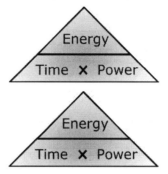

10.2) An electrical appliance converts 360,000 joules of electrical energy every hour when it is switched on. Calculate its power rating.

power = $\dfrac{\text{energy}}{\text{time}}$

= $\dfrac{360{,}000 \text{ J}}{60 \text{ seconds per minute} \times 60 \text{ minutes per hour}}$

= 100 W

10.3) How much energy does a 60 W light bulb that is turned on for 3 hours a day use in one year?

energy = time x power
      = 60 seconds per minute x 60 minutes per hour x 3 hours per day x 365 days per year x 60 W
      = 236,520,000 J
      = 237 MJ

10.4) What is the difference between a step up transformer and a step down transformer? Use this information to comment on the number of windings on a transformer.

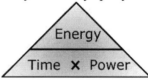

$$\dfrac{V_p}{V_s} = \dfrac{N_p}{N_s}$$

A step up transformer takes the input voltage and gives a higher output voltage (it makes the voltage bigger).

A step down transformer takes the input voltage and gives a lower output voltage.

Using the equation for a transformer, if the primary voltage is smaller than the secondary voltage then there must be more turns on the secondary coil than the primary coil.

If the primary voltage is larger than the secondary voltage, then there must be more turns on the primary coil than the secondary coil.

(More voltage in than out → more turns on the way in than the way out, which means more turns on the primary coil than on the secondary coil.

More voltage out than in → more turns on the way out, which means more turns on the secondary coil than the primary coil.)

10.5) I need to design a transformer that will allow me to get 12 volts from a 230 volt power supply. Suggest an appropriate number of turns.

$$\frac{230 \text{ V}}{12 \text{ V}} = \frac{N_p}{N_s}$$

$$\boxed{\frac{V_p}{V_s} = \frac{N_p}{N_s}}$$

N_p = 230 turns
N_s = 12 turns

Or any number that maintains this ratio, i.e. N_p = 115 and N_s = 6 or N_p = 460 and N_s = 24 etc.

10.6) A transformer has 2 turns on its primary coil and 2,000 turns on its secondary coil. The input voltage is 230 volts. Calculate the output voltage.

$$V_s = \frac{N_s \times V_p}{N_p}$$

$$\boxed{\frac{V_p \times N_s}{V_s \times N_p}}$$

$$= \frac{2,000 \text{ turns} \times 230 \text{ V}}{2 \text{ turns}}$$

= 230,000 volts
= 230 kV

10.7) A transformer has an output voltage of 10,000 volts. It has an input voltage of 100,000 volts. The number of turns on the primary coil is 5,000. How many turns are there on the secondary coil?

$$N_s = \frac{N_p \times V_s}{V_p}$$

$$\boxed{\frac{V_p \times N_s}{V_s \times N_p}}$$

$$= \frac{5,000 \text{ turns} \times 10,000 \text{ V}}{100,000 \text{ V}}$$

= 500 turns

10.8) A voltage of 10,000 volts is changed down to a voltage of 5,000 volts using a transformer. The power supplied to the transformer at the primary coil is 1,000 watts. The electricity is then transmitted along a cable of resistance 2 ohms.
   a) Calculate the power that is lost in the cable.
   b) Calculate the power that would have been lost in the cable if the transformer was connected the other way around.
   c) Use this answer to comment on why electricity is transported across the country at high voltage.

a) Step 1: Calculate the current on the output side.

$$\text{current} = \frac{\text{power}}{\text{voltage}}$$
$$= \frac{1,000 \text{ W}}{5,000 \text{ V}}$$
$$= 0.2 \text{ amps}$$

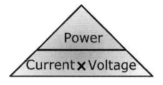

Step 2: Calculate the power lost in the cable.

power = current x current x resistance
= 0.2 amps x 0.2 amps x 2 ohms
= 0.08 watts

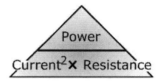

b) Step 1: Find the number of coils on the transformer. To do this we need to suggest a number of turns for the primary and secondary coil. This is needed to calculate a value for which the voltage would be changed to. For the setup given in the question, 10,000 V is changed to 5,000 V.

$$\frac{V_p}{V_s} = \frac{10,000 \text{ V}}{5,000 \text{ V}}$$

$$\frac{V_p}{V_s} = \frac{N_p}{N_s}$$

From part 1

$$\frac{10,000}{5,000} = \frac{N_p}{N_s}$$

A suggested number of turns for the transformer is N_p = 10,000 turns and N_s = 5,000 turns, but any ratio of this form would work, i.e. N_p = 10 and N_s = 5 or N_p = 1000 and N_s = 500 etc.

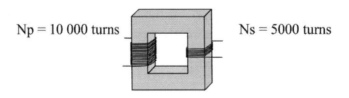

Transformer connected in original direction

If the transformer is now connected the other way around then the number of turns in N_p becomes 5,000 turns and N_s becomes 10,000 turns. Remember also that V_p is fixed by the question at 10,000 V and the power in and out is given as 1,000 W.

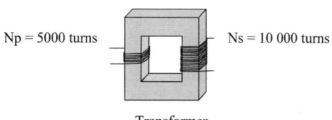

Transformer connected in other direction

Step 2: Calculate the output voltage and from there the output current.

$$V_s = \frac{N_s \times V_p}{N_p}$$

$$= \frac{5,000 \text{ turns} \times 10,000 \text{ V}}{10,000 \text{ turns}}$$

$$= 20,000 \text{ volts}$$

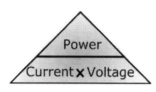

$$\text{current} = \frac{\text{power}}{\text{voltage}}$$

$$= \frac{1,000 \text{ W}}{20,000 \text{ V}}$$

$$= 0.05 \text{ amps}$$

Step 3: Use the value of the output current and the cable resistance to calculate the power lost to heat in the cable.

power = current x current x resistance
power lost as heat = 0.05 amps x 0.05 amps x 2 ohms
= 0.005 watts

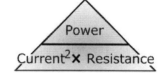

c) Power is transmitted at higher voltage to reduce energy loss as the power lost is proportional to the square of the current.

10.9) There is a supply voltage of 230 volts at the primary coil of my transformer. The output coil supplies a voltage of 230,000 volts. If the input current is 10 amps, calculate the value of the output current.

power in = power out

$$I_{out} = \frac{I_{in} \times V_{in}}{V_{out}}$$

$$= \frac{230 \text{ V} \times 10 \text{ A}}{230,000 \text{ V}}$$

$$\boxed{\frac{Current_{in} \times Voltage_{in}}{Current_{out} \times Voltage_{out}}}$$

current out = 0.01 amps

10.10) 2 transformers are connected in series. A step down transformer is followed by a step up transformer. The step down transformer has 2,000 windings on the primary coil and 250 windings on the secondary coil. The step up transformer has 400 windings on the primary coil and 40,000 on the secondary coil. A supply voltage of 40 volts is fed into the primary coil of the step down transformer. It is then fed directly from the output of the step down transformer to the input of the step up transformer.

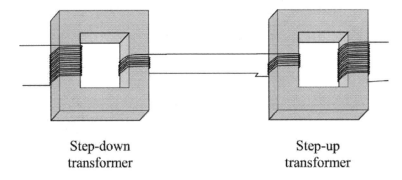

Step-down transformer   Step-up transformer

Calculate the final output voltage from the step up transformer.

Step 1: Find the output voltage for transformer 1. This will become the input voltage for transformer 2.

$$V_s = \frac{N_s \times V_p}{N_p}$$

$$= \frac{250 \text{ turns} \times 40 \text{ V}}{2,000 \text{ turns}}$$

$$\boxed{\frac{V_p \times N_s}{V_s \times N_p}}$$

= 5 volts

Step 2: Calculate the output voltage for transformer 2 (using V_s for transformer 1 as the input).

$$V_s = \frac{N_s \times V_p}{N_p}$$

$$= \frac{40,000 \text{ turns} \times 5 \text{ V}}{400 \text{ turns}}$$

$$\boxed{\frac{V_p \times N_s}{V_s \times N_p}}$$

= 500 volts

The final output voltage from the second transformer is 500 volts.

Bonus Questions 10.1: Electrical Power 2 and Transformers

10.1.1) A 12 V electricity supply is connected across a bulb of resistance 1 Ω, a bulb of resistance 2 Ω and a bulb of resistance 3 Ω all in series. What currents do each of the bulbs draw from the power supply? How much power do each of the bulbs transform from electrical power to heat and light?

10.1.2) A 100 W bulb draws a current of 10 amps when it is connected to its voltage supply. Calculate the value of the resistance in the bulb.

10.1.3) What is the current in a 12 W resistor that has a resistance of 10 Ω?

10.1.4) Calculate the power rating for a circuit that is comprised of four 3 Ω resistors in parallel that is being supplied by a 2 amp current.

10.1.5) A 50 W bulb is supplied by a 110 V power supply. Calculate the value of the resistance of the bulb.

10.1.6) Calculate the power dissipated by a resistance of 3 Ω that has a voltage of 20 V across it.

10.1.7) What is the power rating of a circuit with resistance 30 Ω that is designed to work with a 50 V power supply?

10.1.8) Four 50 W light bulbs designed to work with a voltage of 12 V are connected in parallel. A 12 V power supply is then connected across the circuit. What is the total power output of the lights?

10.1.9) In the question above, what would happen if the lights were connected in series, rather than in parallel.

10.1.10) A current of 12 amps is supplied to a circuit by a supply voltage of 24 V. Calculate the power dissipated by the circuit.

10.1.11) What voltage is required to supply a 40 W bulb with a current of 2 amps?

10.1.12) Calculate the voltage that should be applied across a circuit with a power rating of 50 W and a current rating of 4 amps.

10.1.13) A circuit containing a 4 Ω resistance is supplied by a 20 V power supply. Calculate the value of the power that the circuit dissipates.

10.1.14) Calculate the output voltage of a transformer that is supplied with an input voltage of 50 volts and has 50 turns on the primary coil and 350 turns on the output coil.

10.1.15) What input voltage must be supplied to get an output voltage of 300 V from a transformer with 20 turns on the input coil and 100 coils on the output coil?

10.1.16) Suggest a good number of coils to have on a transformer that can convert 115 volts to 12 volts.

10.1.17) A lighting circuit is supposed to be powered by a 12 V power supply. This comes from a step down transformer that converts 110 V to 12 V. The transformer was accidentally installed the wrong way round. When a bulb is connected to the circuit the bulb breaks.
   a) Explain why it broke.
   b) Calculate the voltage that was actually supplied to the 12 V bulb.

10.1.18) A transformer has a primary coil containing 40 coils and a secondary coil containing 320 coils. If the primary coil is supplied with a direct current of 10 volts how many volts are present in the secondary coil?

10.1.19) A power station transmits electrical energy at a very high voltage with a current of 10 A. This passes through several kilometres of electrical cables with a total combined resistance of 4 Ω.
   a) Calculate the power lost due to the electrical cables.
   b) Comment on why large factories are often built near power stations.

10.1.20) For our calculations we assume that transformers are 100% efficient. Calculate the output current from a step down transformer that has an output of 12 V from an input of 120 V if the input current is 0.5 amps and the transformer is only 95% efficient.

Bonus Questions 10.2: Electrical Power 2 and Transformers

10.2.1) A 24 V electricity supply is connected across a heater of resistance 4 Ω, a heater of resistance 8 Ω and a heater of resistance 12 Ω all in series. What currents do each of the heaters draw from the power supply? How much electrical power do each of the heaters transform?

10.2.2) A 300 W amplifier draws a current of 5 amps when it is connected to its voltage supply. Calculate the value of the resistance in the amplifier.

10.2.3) What is the operating current in a 50 W bulb that has a resistance of 10 Ω?

10.2.4) Calculate the power rating for a circuit that is comprised of three 6 Ω resistors in parallel that is being supplied by a 4 amp current.

10.2.5) A 150 W light bulb is supplied by a 110 V power supply. Calculate the value of the resistance of the bulb.

10.2.6) What is the power dissipated by a resistance of 16 Ω that has a voltage of 20 V across it?

10.2.7) Calculate the power rating of a circuit of resistance 600 Ω that is designed to work with a 15 V power supply.

10.2.8) A 100 W light, a 400 W heater and a 10 W fan are all designed to work with a 12 V power supply. They are connected in parallel and a 12 V power supply is then connected across the circuit. What is the total power output of the arrangement?

10.2.9) In the question above, what would happen to the total power output if the components were connected in series, rather than in parallel?

10.2.10) A supply voltage of 50 V delivers a current of 6 amps to a circuit. Calculate the power dissipated by the circuit.

10.2.11) What voltage is required to supply an 8 W bulb with a current of 0.73 amps?

10.2.12) How much voltage should be applied across a circuit with a power rating of 90 W and a current rating of 2 amps?

10.2.13) Calculate the value of the power dissipated by a circuit containing a 10 Ω resistance supplied by a 24 V power supply.

10.2.14) What is the value of the output voltage of a transformer that is supplied with an input voltage of 200 volts and has 30 turns on the primary coil and 420 turns on the output coil?

10.2.15) Calculate the value of the input voltage required to get an output voltage of 150 V from a transformer with 250 turns on the input coil and 100 coils on the output coil.

10.2.16) Suggest a good number of coils to have on a transformer that can convert 230 volts to 12 volts.

10.2.17) A microchip circuit is supposed to be powered by a 12 V supply. This comes from a step down transformer that converts 230 V to 12 V. The transformer was accidentally installed the wrong way round. When the setup was connected to the power supply the circuit broke.
   a) Why did it break?
   b) How much voltage was actually supplied to the 12 V bulb?

10.2.18) A transformer has a primary coil containing 32 coils and a secondary coil containing 4,800 coils. If the primary coil is supplied with a direct current of 10 volts how many volts are present in the secondary coil?

10.2.19) A power station transmits electrical energy at a very high voltage with a current of 15 A. This passes through several kilometres of electrical cables with a total combined resistance of 2 Ω.
   a) Calculate the energy lost due to the electrical cables.
   b) Comment on why electrical cables on the overhead electric lines are so thick even though it would make the lines heavier and make the pylons more susceptible to damage from heavy snow or ice.

10.2.20) What would the output be if the following transformer had an efficiency of only 98%? (For our calculations we normally assume that transformers are 100% efficient.) Calculate the output current from a step up transformer that has an output of 1,100 V from an input of 25 V if the input current is 200 amps and the transformer is only 95% efficient.

11. Waves, Light and Colours

Silly New Rules

<u>S</u>illy <u>**new**</u> <u>r</u>ules.
<u>S</u>ome <u>f</u>ed up <u>w</u>orkers
<u>n</u>ever <u>**sign**</u>ed <u>**in**</u>, <u>n</u>ever <u>**sign**</u>ed <u>**out.**</u>

<u>S</u>illy <u>new</u> <u>r</u>ules.
<u>S</u>ome <u>f</u>ed up <u>w</u>orkers
<u>n</u>ever <u>sign</u>ed <u>in</u>, <u>n</u>ever <u>sign</u>ed <u>out</u>.

<u>S</u>illy <u>n</u>ew <u>r</u>ules.	s n* x r	
<u>S</u>ome <u>f</u>ed up <u>w</u>orkers	s f x w	
<u>N</u>ever <u>sign</u>ed <u>in</u> <u>n</u>ever <u>sign</u>ed <u>out</u>.	$n_1 \sin \theta_1$ = $n_2 \sin \theta_2$	$\dfrac{n_1 \sin \theta_1}{n_2 \sin \theta_2}$

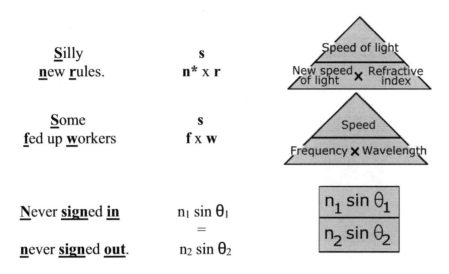

The refractive index of air (n_{air}) has a value of 1, as does a vacuum like space. (This is not exactly true but it's so close that, at this level, we use the value 1.)

* Just a note about the memory aid here. I have written s = new x r. Here r stands for refractive index and new is the new speed of light. This is purely to help you remember. If you are to write this in an exam use the full equation. This is because, as you can see in the first equation, n has a very specific meaning in waves. It is the symbol that is used for the refractive index.

So the full equation here would really be:

speed of light = n x new speed of light in the medium

Writing the equations in full when you answer questions is an essential part to getting the full marks.

This then brings us to our final three statements to help us remember our physics. The first thing that we will remember tells us about all of the different types of electromagnetic wave, most of which we can't see with our eyes and need equipment to see for us.

In the second part we will look at how lenses work with a short journey and in the third part we will look at all of the different types of colours that we can see with our eyes.

Short Gavin

Short Gavin e**x**ecutes **u**nwelcome **v**isitors **i**n a **m**ad **r**age.

So let us begin with the all of the types of electromagnetic wave.

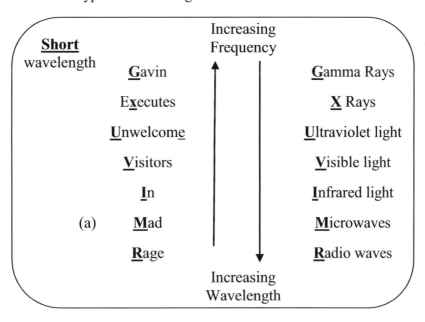

It is important to notice that as the wavelength is increasing here the frequency is decreasing. When the frequency gets bigger the wavelength gets smaller. This is because when you multiply them together the answer that you get is the speed of light 3×10^8 ms^{-1} or 300,000,000 ms^{-1}.

Which Path?

If you were a light ray going through a lens, you would stand at the top of an object and think. Which of the 3 directions would you choose?

1. Would you go the fastest route possible right through the centre of the lens?

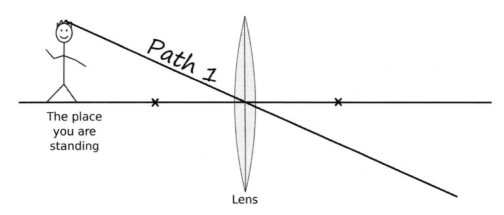

2. Would you go the fastest route possible to the lens and then visit the interesting focal point?

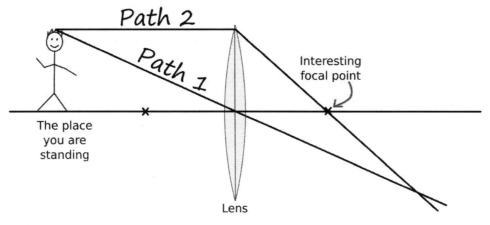

3. Would want to take your time to get to the lens visiting the interesting focal point first and then get to your destination as fast as possible?

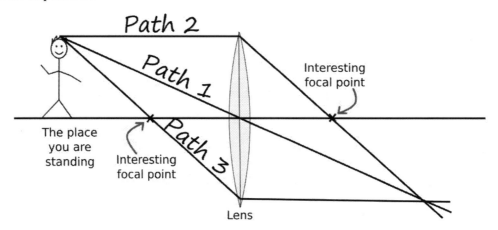

When light goes through a simple converging lens all you need to remember is that there are three different paths you can take. Where all three paths intersect is the top of the image of the object at the start (now it's upside down).

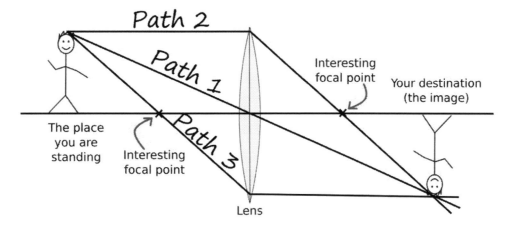

This is exactly how we approach drawing the images created with lenses.

There are 3 straight lines all drawn from the top of the object. 2 of these lines can change direction and 1 does not.
- The first travels through the centre of the lens. It does not change direction.
- The second goes through the focal point on the left until the middle of the lens is reached. Then the direction changes and it travels straight on to the right.
- The third line travels straight to the right and then, once it reaches the middle of the lens, the direction changes and it travels straight through the focal point on the right of the lens.

So you would either draw, or be given an image like the following and be asked to show where the image is created and describe it. For this you need to know where the focal points are (these are normally given in the question).

The x on either side of the lens is the focal point. It is where rays coming in parallel to each other and parallel to the long horizontal line would meet. It is a distance of 1 focal length from the centre of the lens. These are often labelled with an f or f1 and f2. The distances between the lens and f1 and the lens and f2 are the same.

Example: complete the diagram:

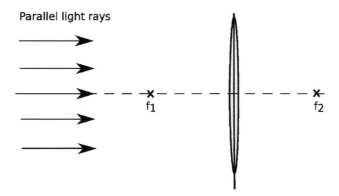

Answer:

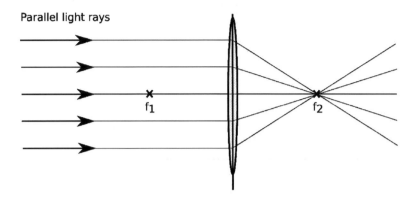

If in doubt a good rule for dealing with parallel rays is that the line going through the very centre of the lens is always unchanged. The lines will all come together when they are a distance of 1 focal length from the lens.

Example: An image is located a distance of 2 times the focal length from a thin converging lens. Complete the ray diagram and show the final location of the image. Draw and label the image.

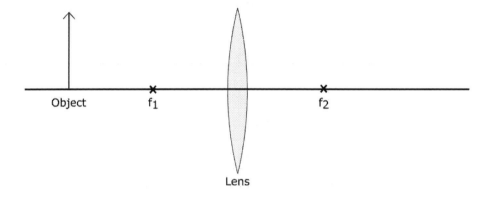

Drawing a line through the centre of the lens we can use this as the point where the straight lines can turn. Then we draw our straight lines as before and, if the lines meet on the right hand side (they often do) then we draw the image (which will look the same as the object) upside down and scale it so that it fits.

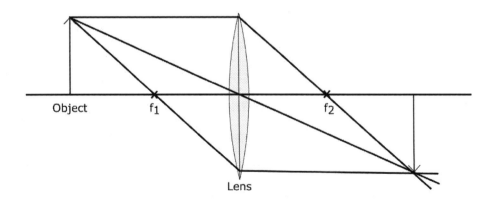

Real images: If the image is located on the opposite side to the object then it is called a real image. These images can be projected and they are how cinema's project films. If it is upside down, then it is called an inverted image.

Drawing these 3 lines in this way will give you the location of the final image.

Final part to be aware of:

There are 2 circumstances where the 3 lines do not come together.
- When the object is at the focal length an image is not formed.
- When an object is between the lens and the focal point an image is formed but it is on the same side of the lens as the object. It is called a virtual image and it cannot be projected. It is what you see when you use a magnifying glass on a newspaper.

Virtual Images: We find this when we try to draw the 3 lines. One of the lines will disappear into the distance, the other two will move farther apart and the only way to bring them together is to continue the line backwards past the object so that they meet behind the object. This image is called a virtual image because it can't be projected. It is also the right way up (not inverted) and it is larger than the original object – which is why we can use a magnifying glass to see things that are small.

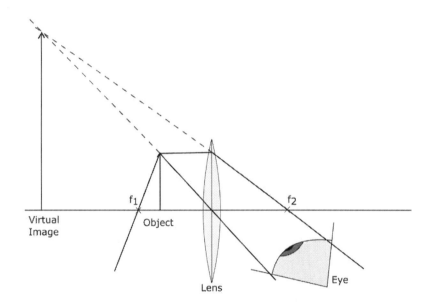

A virtual image using a magnifying glass:

Orange Rabbits

We also need to remember the colours of visible light!

<u>V</u>ery <u>i</u>nteresting <u>b</u>unnies <u>g</u>ive <u>y</u>ou <u>o</u>range <u>r</u>abbits

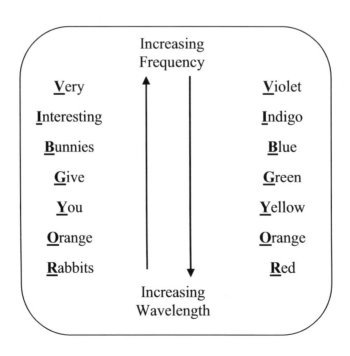

Examples: Waves, Light and Colours

Example 1) I have a wave of wavelength 50 metres and frequency 3 Hz. What is the speed of the wave?

speed = frequency x wavelength
    = 3 Hz x 50 m
    = 150 ms^{-1}

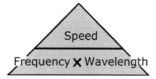

Example 2) A wave has a speed of 340 ms^{-1} in a medium. It has a frequency of 34 Hz. Calculate the wavelength of the wave.

wavelength = $\dfrac{\text{speed}}{\text{frequency}}$

           = $\dfrac{340 \text{ ms}^{-1}}{34 \text{ Hz}}$

           = 10 m

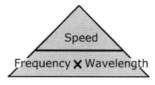

Example 3) A wave has a speed of 1 kms^{-1} and a wavelength of 25 m. Calculate the frequency of the wave.

frequency = $\dfrac{\text{speed}}{\text{wavelength}}$

          = $\dfrac{1{,}000 \text{ ms}^{-1}}{25 \text{ m}}$

          = 40 Hz

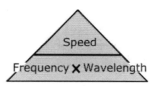

Example 4) The refractive index of a medium is 1.5. Calculate the new speed of light in this medium.

new speed of light = speed of light / refractive index
                     = 3 x 10^8 ms^{-1} / 1.5
                     = 2 x 10^8 ms^{-1}

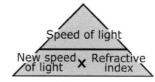

Example 5) The new speed of light in a medium is 150,000,000 ms^{-1}. Calculate the value of the refractive index of the medium.

refractive index = $\dfrac{\text{speed of light}}{\text{new speed of light}}$

               = $\dfrac{3 \times 10^8 \text{ ms}^{-1}}{1.5 \times 10^8 \text{ ms}^{-1}}$

               = 2

Example 6) The refractive index of a material is 4.0 and the speed of light in the material is 75,000 kms^{-1}. Calculate the speed of light in a vacuum.

speed of light = new speed of light x refractive index
= 75 x 10^6 ms^{-1} x 4
= 3 x 10^8 ms^{-1}

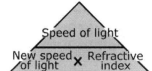

Example 7) The speed of light changes in different materials. I have 3 materials one of refractive index 1.5 (glass), one of refractive index 1 (air) and one of refractive index 1.7 (lead glass).
a) Calculate the speed of light in those different materials.

$$\text{new speed of light} = \frac{\text{speed of light}}{\text{refractive index}}$$

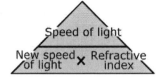

Glass

new speed of light = $\frac{3 \times 10^8}{1.5}$

= 2 x 10^8 ms^{-1}

Air

new speed of light = $\frac{3 \times 10^8}{1}$

= 3 x 10^8 ms^{-1}

Lead Glass

new speed of light = $\frac{3 \times 10^8}{1.7}$

= 1.75 x 10^8 ms^{-1}

b) Using the information above and also that the frequency of blue light in a vacuum is 6.7 x 10^{14} Hz, calculate the wavelength of blue light in space and in each of the 3 mediums in the question.

$$\text{wavelength} = \frac{\text{speed}}{\text{frequency}}$$

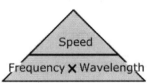

In a vacuum (space is a vacuum!)

wavelength = $\frac{\text{speed}}{\text{frequency}}$

= $\frac{3 \times 10^8 \text{ ms}^{-1}}{6.7 \times 10^{14} \text{ Hz}}$

= 447 x 10^{-9} m
= 447 nm

In air

$$\text{wavelength} = \frac{3 \times 10^8 \text{ ms}^{-1}}{6.7 \times 10^{14} \text{ Hz}}$$

$$= 447 \times 10^{-9} \text{ m}$$
$$= 447 \text{ nm}$$

In Glass

$$\text{wavelength} = \frac{2 \times 10^8 \text{ ms}^{-1}}{6.7 \times 10^{14} \text{ Hz}}$$

$$= 299 \times 10^{-9} \text{ m}$$
$$= 299 \text{ nm}$$

In Lead Glass

$$\text{wavelength} = \frac{1.75 \times 10^8 \text{ ms}^{-1}}{6.7 \times 10^{14} \text{ Hz}}$$

$$= 261 \times 10^{-9} \text{ m}$$
$$= 261 \text{ nm}$$

Example 8) Some scientists measure the refractive index of a material to be 0.8. Is this possible?

$$\text{new speed of light} = \frac{\text{speed of light}}{\text{refractive index}}$$

$$= \frac{3 \times 10^8 \text{ ms}^{-1}}{0.8}$$

$$= 3.75 \times 10^8 \text{ ms}^{-1}$$

Speed of light = New speed of light × Refractive index

This is impossible as nothing can travel faster than the speed of light (when it is in a vacuum). This is one of the rules of the universe. There may be ways to avoid the rule in the distant future but for the moment it appears that the rule itself is unbreakable. This is an exciting area of research. That is why all refractive index values are 1 or more. There is no material with a refractive index less than 1.

Example 9) A ray of light is travelling from air into glass ($n_{air} = 1$, $n_{glass} = 1.5$). It is travelling through the air and strikes the glass at an angle of 25 degrees to the normal line. What angle does the light then travel into the glass with?

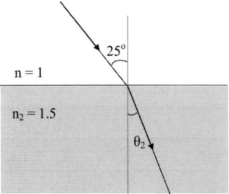

Make sure that your calculator is set to work in degrees!

$$\sin(\theta_2) = \frac{n_1 \times \sin(\theta_1)}{n_2}$$
$$= \frac{1 \times \sin(25°)}{1.5}$$
$$= 0.282$$

$\theta_2 = \sin^{-1}(0.282)$
$ = 16.4°$

$\dfrac{n_1 \sin \theta_1}{n_2 \sin \theta_2}$

Example 10) A ray of light is incident on a block at an angle of 20 degrees. It travels into the block with a new angle of 14 degrees. The original medium was air which has a refractive index of 1. Calculate the value of the refractive index of the new material.

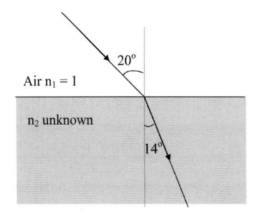

$$n_2 = \frac{n_1 \times \sin(\theta_1)}{\sin(\theta_2)}$$
$$= \frac{1 \times \sin(25°)}{\sin(14°)}$$
$$= 1.75$$

$\dfrac{n_1 \sin \theta_1}{n_2 \sin \theta_2}$

Example 11) An object is placed at a distance of 4 times the focal length away from a thin converging lens. Draw a sketch of the ray diagram and describe the image created.

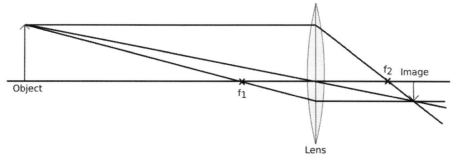

Image is inverted
Image is real
Image is smaller than the object
Image is closer to the lens than the object

Example 12) An object is placed at a distance of 1.5 times the focal length away from a thin converging lens. Draw a ray diagram and describe the image created.

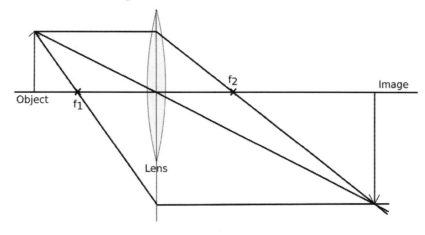

Image is inverted
Image is real
Image is larger than the object
Image is farther away from the lens than the object

Example 13) An object is placed a distance of 2 times the focal length away from a thin converging lens. Describe the image created.

Image is inverted
Image is real
Image is the same size as the object
Image is the same distance from the lens than the object

Questions: Waves, Light and Colours

11.1) The speed of a wave in water is dependent on how deep the water is. The waves on a shallow puddle travel at a speed of 10 cms^{-1}. They have a wavelength of 2 cm. Calculate the frequency of the wave.

11.2) A wave has a frequency of 30 Hz and a wavelength of 11 m. Calculate the speed of the wave.

11.3) The human ear can hear frequencies of between 20 Hz and 20,000 Hz. Calculate the corresponding wavelengths.
The speed of sound in air is 340 ms^{-1}.

11.4) The speed of sound underwater is about 1,500 ms^{-1}. Explain why this makes it difficult for a person to tell the direction of where a sound comes from.

11.5) Use questions 3 and 4 to find the maximum and minimum wavelengths that the human ear can detect underwater.

11.6) Bats use sonar to tell where their prey is. They release a chirp and then wait for the echo (when the sound bounces off their prey and returns to the bat). A bat releases a chirp and then hears the echo 0.2 seconds later. The speed of sound in air is 340 ms^{-1}. Calculate the distance from the bat to its prey.

11.7) Dolphins also use sonar to find fish. Calculate the distance from the dolphin to some fish if it hears the echo after 0.1 seconds. The speed of sound in water is 1,500 ms^{-1}.

11.8) Calculate the speed with which light will travel in materials of the following refractive indices.
   a) 1.2
   b) 1.4
   c) 1.9
   d) 2.0
   e) 0.5
   f) Is it possible to have a material with the refractive index listed in e)? Explain your answer.

11.9) The speed of light in a material is 1.75×10^8 ms^{-1}. Calculate its refractive index.

11.10) In a vacuum light has a wavelength of 510 nm. The light travels into a glass block of refractive index 3. Calculate the new speed, frequency and wavelength of the light.

11.11) Calculate the angle with which a ray of light will continue into a block of glass of refractive index 1.3. Its angle of incidence is 35 degrees and it is travelling into the block from air. $n_{air} = 1$

11.12) Light enters a glass block with refractive index 2. Its angle of refraction inside the glass block is 12 degrees. Assuming that it entered from air calculate the angle of incidence to the glass block. $n_{air} = 1$

11.13) An object is placed at a distance of 3 times the focal length away from a thin converging lens. Draw a ray diagram and describe the image created.

11.14) An object is placed a distance of 0.5 times the focal length away from a thin converging lens. Draw a ray diagram and describe the image created.

11.15) An object is placed a distance of 2.5 times the focal length away from a thin converging lens. Draw a ray diagram and describe the image created.

11.16) An image is created using a thin converging lens. The image is twice the size of the object. It is inverted and it is real. Describe the position of the object in relation to the lens.

11.17) Place the following in order of danger, from most to least dangerous.
        Radio waves
        Gamma Rays
        X-rays
        Visible light

11.18) List uses for the following: X-rays, UV light, sunlight, microwaves and radio waves.

11.19) Place the following in order of their;
    a) Wavelengths (ascending)
    b) Frequency (ascending)
    c) Energy (ascending)
Gamma rays, UV light, Visible light, Radio waves, Microwaves, X-Rays, Infrared light

Answers: Waves, Light and Colours

*The speed of sound in air changes depending mostly on temperature but also, to a lesser extent, pressure and humidity. Different values may be given to you depending in the question that you are given. The most common value used is 340 ms^{-1} however it is also possible to have a value of 330 ms^{-1}. Because of this it is important to pay attention to the value that you need to use for your exams. Your teacher may give you this or it may be written on the exam if you need it. The value I wish you to use in these questions is 340 ms^{-1}.

11.1) The speed of a wave in water is dependent on how deep the water is. The waves on a shallow puddle travel at a speed of 10 cms^{-1}. They have a wavelength of 2 cm. Calculate the frequency of the wave.

$$\text{frequency} = \frac{\text{speed}}{\text{wavelength}}$$
$$= \frac{0.1 \text{ ms}^{-1}}{0.02 \text{ m}}$$
$$= 5 \text{ Hz}$$

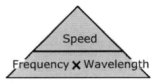

11.2) A wave has a frequency of 30 Hz and a wavelength of 11 m. Calculate the speed of the wave.

speed = frequency x wavelength
= 30 Hz x 11 m
= 330 ms^{-1}

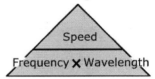

11.3) The human ear can hear frequencies of between 20 Hz and 20,000 Hz. Calculate the corresponding wavelengths. The speed of sound in air is 340 ms^{-1}.

At 20Hz

$$\text{wavelength} = \frac{\text{speed}}{\text{frequency}}$$
$$= \frac{340 \text{ ms}^{-1}}{20 \text{ Hz}}$$
$$= 17 \text{ m}$$

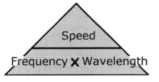

At 20,000 Hz

$$\text{wavelength} = \frac{\text{speed}}{\text{wavelength}}$$
$$= \frac{340 \text{ ms}^{-1}}{20,000 \text{ Hz}}$$
$$= 0.017 \text{ m or } 1.7 \text{ cm.}$$

The human ear can hear wavelengths between 1.7 cm and 17 m.

11.4) The speed of sound underwater is about 1,500 ms⁻¹. Explain why this makes it difficult for a person to tell the direction of where a sound comes from.

The human brain uses the time delay from the sounds reaching each ear to tell direction. If the time delay is too small people are unable to tell the direction that a sound comes from. As the sound travels so quickly underwater it is not possible for people to tell the direction that it is coming from.

11.5) Use questions 3 and 4 to find the maximum and minimum wavelength that the human ear can detect underwater.

At 20 Hz

wavelength = $\dfrac{\text{speed}}{\text{frequency}}$

maximum wavelength = $\dfrac{1{,}500 \text{ ms}^{-1}}{20 \text{ Hz}}$

= 75 m

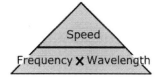

At 20, 000 Hz

minimum wavelength = $\dfrac{1{,}500 \text{ ms}^{-1}}{20{,}000 \text{ Hz}}$

= 0.075 m

The human ear can hear wavelengths between 0.075 m (7.5 cm) and 75 m underwater.

11.6) Bats use sonar to tell where their prey is. They release a chirp and then wait for the echo (when the sound bounces off their prey and returns to the bat). A bat releases a chirp and then hears the echo 0.2 seconds later. The speed of sound in air is 340 ms⁻¹. Calculate the distance from the bat to its prey.

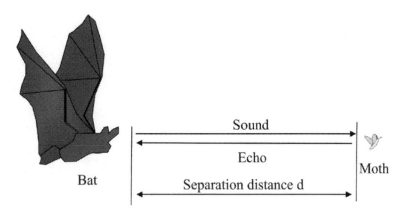

The sound has to travel from the bat to its prey (this is a distance d) and then travel back again (another distance d) so that the bat can hear the echo. This means that the sound has to travel a total distance of 2d.

2d = time x speed
   = 0.2 s x 340 ms⁻¹
   = 68 m
d = 34 m

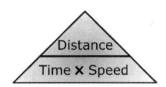

The distance from the bat to its prey is 34 m.

11.7) Dolphins also use sonar to find fish. Calculate the distance from the dolphin to some fish if it hears the echo after 0.1 seconds. The speed of sound in water is 1,500 ms⁻¹.

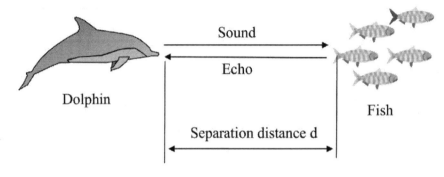

The sound has to travel from the dolphin to the fish and then travel back again so that the dolphin can hear the echo. This means that the sound has to travel a total distance of 2d.

2d = time x speed
   = 0.1 s x 1,500 ms⁻¹
   = 150 m
d = 75 m

The distance from the dolphin to the fish is 75 m.

11.8) Calculate the speed with which light will travel in materials of the following refractive indices.
   a) 1.2
   b) 1.4
   c) 1.9
   d) 2.0
   e) 0.5
   f) Is it possible to have a material with the refractive index listed in e)? Explain your answer.

new speed of light = $\dfrac{\text{speed of light}}{\text{refractive index}}$

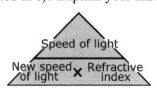

This equation will be required for parts a to e of this question.

a) new speed of light = $\dfrac{3 \times 10^8 \text{ ms}^{-1}}{1.2}$

   = 2.5×10^8 ms⁻¹

242

b) new speed of light $= \dfrac{3 \times 10^8 \text{ ms}^{-1}}{1.4}$

$= 2.14 \times 10^8 \text{ ms}^{-1}$

c) new speed of light $= \dfrac{3 \times 10^8 \text{ ms}^{-1}}{1.9}$

$= 1.58 \times 10^8 \text{ ms}^{-1}$

d) new speed of light $= \dfrac{3 \times 10^8 \text{ ms}^{-1}}{2.0}$

$= 1.5 \times 10^8 \text{ ms}^{-1}$

e) new speed of light $= \dfrac{3 \times 10^8 \text{ ms}^{-1}}{0.5}$

$= 6 \times 10^8 \text{ ms}^{-1}$

f) It is not possible for a material to have a refractive index of 0.5. Nothing can travel faster than the speed of light in a vacuum. The refractive index must be 1 or higher.

11.9) The speed of light in a material is 1.75×10^8 ms^{-1}. Calculate its refractive index.

refractive index $= \dfrac{\text{speed of light}}{\text{new speed of light}}$

$= \dfrac{3 \times 10^8 \text{ ms}^{-1}}{1.75 \times 10^8 \text{ ms}^{-1}}$

$= 1.71$

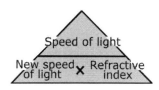

11.10) In a vacuum light has a wavelength of 510 nm. The light travels into a glass block of refractive index 3. Calculate the new speed, frequency and wavelength of the light.

new speed of light $= \dfrac{\text{speed of light}}{\text{refractive index}}$

$= \dfrac{3 \times 10^8 \text{ ms}^{-1}}{3}$

$= 1 \times 10^8 \text{ ms}^{-1}$

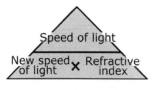

frequency $= \dfrac{\text{speed}}{\text{wavelength}}$

$= \dfrac{3 \times 10^8 \text{ ms}^{-1}}{510 \times 10^{-9} \text{ m}}$

$= 5.88 \times 10^{14}$ Hz

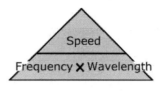

Frequency of light is not affected by changing mediums (it never changes). The frequency remains 5.88×10^{14} Hz.

wavelength $= \dfrac{\text{speed}}{\text{frequency}}$

$= \dfrac{1 \times 10^8 \text{ ms}^{-1}}{5.88 \times 10^{14} \text{ Hz}}$

$= 1.7 \times 10^{-7}$ m

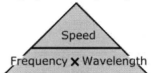

11.11) Calculate the angle with which a ray of light will continue into a block of glass of refractive index 1.3. Its angle of incidence is 35 degrees and it is travelling into the block from air. The refractive index of air, $n_{air} = 1$

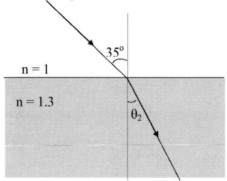

$\sin(\theta_2) = \dfrac{n_1 \times \sin(\theta_1)}{n_2}$

$= \dfrac{1 \times \sin(35°)}{1.3}$

$= 0.441$

$\theta_2 = \sin^{-1}(0.441)$
$= 26.2°$

$n_1 \sin \theta_1$
$n_2 \sin \theta_2$

11.12) Light enters a glass block with refractive index 2. Its angle of refraction inside the glass block is 12 degrees. Assuming that it entered from air calculate the angle of incidence to the glass block. $n_{air} = 1$

$$\sin(\theta_1) = \frac{n_2 \times \sin(\theta_2)}{n_1}$$

$$= \frac{2 \times \sin(12°)}{1}$$

$$= 0.416$$

$\theta_1 = \sin^{-1}(0.416)$
   $= 24.6$ degrees

$n_1 \sin \theta_1$
$n_2 \sin \theta_2$

11.13) An object is placed at a distance of 3 times the focal length away from a thin converging lens. Draw a ray diagram and describe the image created.

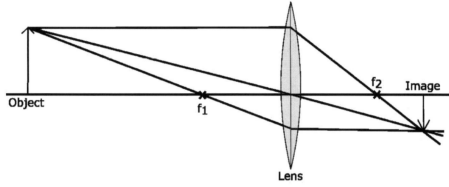

Image is inverted
Image is real
Image is the smaller than the object
Image is closer to the lens than the object

11.14) An object is placed a distance of 0.5 times the focal length away from a thin converging lens. Draw a ray diagram and describe the image created.

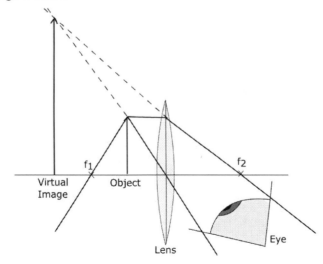

Image is not inverted
Image is not real. It is virtual
Image is larger than the object
Image is farther away from the lens than the object

11.15) An object is placed a distance of 2.5 times the focal length away from a thin converging lens. Draw a ray diagram and describe the image created.

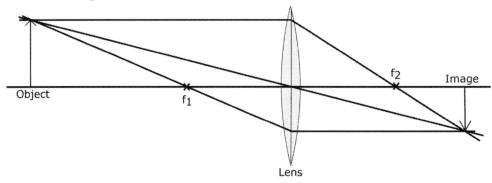

Image is inverted
Image is real
Image is smaller than the object
Image is closer to the lens than the object

11.16) An image is created using a thin converging lens. The image is twice the size of the object. It is inverted and it is real. Describe the position of the object in relation to the lens.

The object is between 1 and 2 focal lengths distance from the lens.

11.17) Place the following in order of danger, from most to least dangerous.
    Radio waves
    Gamma rays
    X-rays
    Visible light

The most dangerous (most energy) are gamma rays. Then X-rays and then visible light. Radio waves are the least dangerous (least energy).

11.18) List uses for the following: X-rays, UV light, visible light, microwaves and radio waves.

X-rays: examining bones in the body without surgery.
UV light: can be used for sunbathing and also to sterilize equipment.
Visible light: we use this to see.
Microwaves: used to cook food and also used by mobile telephones in telecommunication.
Radio waves: used in telecommunication to transmit information (radio).

11.19) Place the following in order of their;
   a) Wavelengths (ascending)
   b) Frequency (ascending)
   c) Energy (ascending)

Gamma rays, UV light, Visible light, Radio waves, Microwaves, X-rays, Infrared light

a) Wavelength:
(smallest) Gamma rays, X-rays, UV light, Visible light, Infrared light, Microwaves, Radio waves (longest)
b) Frequency:
(smallest) Radio waves, Microwaves, Infrared light, Visible light, UV light, X-rays, Gamma rays (highest)
c) Energy:
(smallest) Radio waves, Microwaves, Infrared light, Visible light, UV light, X-rays, Gamma rays (highest)

Bonus Questions 11.1: Waves, Light and Colours

Take the speed of light to be 3×10^8 ms^{-1} and the speed of sound to be 340 ms^{-1} unless otherwise stated in the question.

11.1.1) What is the speed of light in the following media?
   a) Glass: n = 1.4
   b) Plastic: n = 1.7
   c) A vacuum: n = 1

11.1.2) Calculate the frequency of the following types of light, travelling in air (n = 1).
   a) Blue light: wavelength = 470 nm
   b) Green light: wavelength = 510 nm
   c) Red light: wavelength = 650 nm

11.1.3) Certain types of dielectric fluid can be used to slow light to a speed less than 10 ms^{-1}. This allows scientists to see how light behaves when it cannot easily escape. This is similar to behaviour of light in a black hole (but without any dangerous levels of gravity). This type of technique is referred to as an optical black hole. In one of these types of fluid, light travels at 7 ms^{-1}. Calculate the value of the refractive index of this fluid.

11.1.4) Place the following colours in order of wavelength from short to long.
   Red, Orange, Violet, Yellow

11.1.5) Place the following colours in order of frequency from high to low.
   Blue, Violet, Green, Indigo

11.1.6) Write the following colours in order of wavelength from long to short.
   Yellow, Green, Red, Blue

11.1.7) Write down the following colours of light in order of frequency from low to high.
   Indigo, Green, Blue, Orange

11.1.8) List the full spectrum of visible light in ascending frequency order. Once this has been done label at the ends the regions where you would find infrared and ultraviolet light.

11.1.9) Place these types of EM radiation in order of wavelength from short to long:
   Infrared, Radio waves, Gamma rays, Red light, Blue light, UV light

11.1.10) Place the following types of EM radiation in order of how much energy they have from the most energy to least.
   Radio waves, Ultraviolet radiation, X-rays, Infrared, Gamma rays

11.1.11) Describe the essential parts of an EM wave.

11.1.12) A ship sounds its foghorn in the fog. After a time period of 3 seconds an echo of the horn is heard from nearby cliffs. How far away are the cliffs? You may assume the speed of sound in air to be 340 ms^{-1}.

11.1.13) Sound underwater travels at a rate of 1,480 ms^{-1}. A submarine under the surface of the water emits a burst of sonar which bounces off an underwater rock. The echo is heard 6 seconds later by the sonar operator. How far away is the rock?

11.1.14) Assuming that sound travels underwater at a rate of 1,500 ms^{-1} how long would it take for a dolphin, emitting a pulse of sound, to hear an echo from a shoal of fish at a distance of 40 m away?

11.1.15) Light travels from air into water at an angle of 32 degrees to the normal.
Water has a refractive index of 1.33.
    a) Calculate the speed of light in the water.
    b) Calculate the angle of refraction.

11.1.16) Light travels from water with a refractive index of 1.33 into a glass block. The glass has a refractive index of 1.50. If the angle of refraction was 52 degrees, calculate the angle of incidence.

11.1.17) A ray of light travels from air into a block of material. The angle of incidence in air is 68 degrees and the angle of refraction in the block is 50 degrees. Calculate the refractive index of the material. What is the speed of light in the material?

11.1.18) Describe a method to measure the speed of sound.

11.1.19) Light can move from a material with a smaller refractive index into one with a larger refractive index.
    a) In this situation, does the angle that light has to the normal increase, decrease or stay the same?
    b) What about if it travels from a material with a smaller refractive index into one with a larger refractive index?
    c) What happens if the refractive indices are the same?

11.1.20) A student is watching a workman hammer nails into the roof of a house 200 m away. The student sees the hammer hit the nail.
    a) How long does it take before the student hears the sound?
    b) Why does the student hear the sound at a different time to when he sees the hammer hit the nail?

11.1.21) Sound travels at a rate of approximately 1 km per second in liquids and about 5 km per second in solids. How long would it take for sound to travel through 6 km of the following?
    a) Air
    b) Liquid
    c) Rock

11.1.22) When we hear a sound we are able to locate the source of the sound because there is a small but noticeable difference between the time it takes to reach each ear. The brain interprets this and provides a direction. Why is it difficult to tell the direction that a sound comes from when you are underwater?

There are 4 materials each with slightly different properties in the table below. This table will be needed in the next few questions.

	A	B	C	D
Refractive index	1.9	2.7	1.3	5.2
Colour	Light Red	Dark Blue	Light Brown	Dark Green
Density /kgm^{-3}	1,000	700	1,200	2,300
Speed of sound/ms^{-1}	900	1,200	1,500	2,000

11.1.23) In which material does light travel:
    a) The fastest
    b) The slowest

11.1.24) In which material does sound travel:
   a) The fastest
   b) The slowest

11.1.25) A beam of light is sent through each of the materials and the time taken is recorded. Identify each of the following materials.
   a) 1 m of material, time taken = 9×10^{-9} seconds
   b) 10 cm of material, time taken = 1.73×10^{-9}

11.1.26) In another experiment sound is sent through a sample of each of the materials. Identify each of the materials by their results.
   a) 10 cm, time taken = 8.33×10^{-5} seconds
   b) 30 cm, time taken = 1.5×10^{-4} seconds
   c) 15 cm, time taken = 1.0×10^{-4} seconds
   d) 300 cm, time taken = 2.5×10^{-3} seconds

11.1.27) An image is created using a single thin converging lens. The image is twice the size of the object and it is not inverted.
   a) Describe the placement of the object in relation to the lens.
   b) Use this information to decide if the image is real or not.

11.1.28) Complete the diagram:

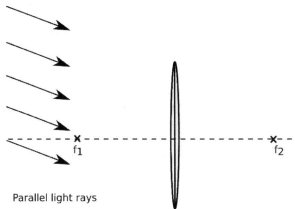

Parallel light rays

11.1.29) Complete the diagram: (Hint, the rules can be applied to either end of the object.)

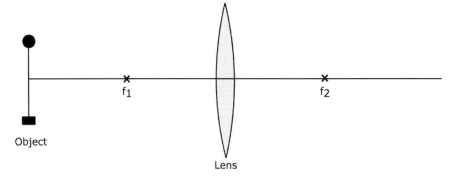

Bonus Questions 11.2: Waves, Light and Colours

Take the speed of light to be 3×10^8 ms^{-1} and the speed of sound to be 340 ms^{-1} unless otherwise stated in the question.

11.2.1) What is the speed of light in the following media?
   a) Polycarbonate: n = 1.6
   b) Sapphire: n = 1.76
   c) A vacuum: n = 1

11.2.2) Calculate the frequency of the following types of light, travelling in air (n = 1).
   a) Orange light: wavelength = 620 nm
   b) Violet light: wavelength = 400 nm
   c) Yellow light: wavelength = 580 nm

11.2.3) There is a rule that says the following: nothing can travel faster than light in a vacuum. Some nuclear reactors are surrounded by water. There are very high energy particles that are travelling close to the speed of light in a vacuum coming out of the nuclear reactor. When they enter the water they give off blue light because they are travelling faster than light in the water (it is the optical equivalent of a sonic boom!). This is known as Cerenkov radiation. The refractive index of the water is 1.3.
   a) Calculate the speed of light in the water.
   b) An electron enters the water with a speed of 2.9×10^8 ms^{-1}. Calculate how much faster the electron was travelling than light in the water.

11.2.4) Place the following colours in order of wavelength from short to long:
   Blue, Violet, Green, Indigo

11.2.5) Place the following colours in order of frequency from high to low:
   Red, Orange, Violet, Yellow

11.2.6) Write the following colours in order of wavelength from long to short:
   Orange, Blue, Yellow, Green

11.2.7) Write down the following colours of light in order of frequency from low to high:
   Blue, Green, Red, Orange

11.2.8) List the full spectrum of visible light in descending frequency order. Once this has been done label at the ends the regions where you would find infrared and ultraviolet light.

11.2.9) Place these types of EM radiation in order of wavelength from low to high:
   X-rays, Radio waves, Gamma rays, Orange light, Yellow light, UV light

11.2.10) Place the following types of EM radiation in order of how much energy they have from the most energy to least:
   Microwaves, Ultraviolet radiation, X-rays, Visible light, Gamma rays

11.2.11) What is an EM wave composed of?

11.2.12) A captain blows a whistle in the fog. After a time period of 6 seconds an echo of the whistle is heard from the edge of a bluff. How far away is the bluff? You may assume the speed of sound in air to be 340 ms^{-1}.

11.2.13) The speed of sound changes according to the temperature of the air. When the air is at 40 degrees Celsius the speed of sound will be about 355 ms^{-1}. When the air is at a temperature of -1 degree Celsius the speed of sound will be 331 ms^{-1}. At the edge of a desert some students are performing an experiment. They set off a firework and time how long it takes to hear an echo from the cliffs 1.5 km away during the daytime when the temperature is 40 degrees Celsius. They repeat the experiment at night time when the temperature is 0 degrees Celsius.
    a) Which measurement takes less time?
    b) How much less time does it take?

11.2.14) Sound travels underwater at a rate of 1,500 ms^{-1} how long would it take for sonar emitted from a boat's fish finder to receive an echo from a shoal of fish at a distance of 16 m away?

11.2.15) Light travels from air into PET plastic at an angle of 36 degrees to the normal. PET has a refractive index of 1.57.
    a) Calculate the speed of light in the PET.
    b) Calculate the angle of refraction.

11.2.16) Light travels from water with a refractive index of 1.33 into a diamond. The diamond has a refractive index of 2.42. If the angle of refraction was 33 degrees, calculate the angle of incidence.

11.2.17) A ray of light travels from a vacuum into a block of material. The angle of incidence from the vacuum is 39 degrees and the angle of refraction in the block is 27 degrees.
    a) Calculate the refractive index of the material.
    b) What is the speed of light in the material?

11.2.18) A pair of students are given a pair of small blocks of wood. They are next to the side of a large building.
    a) What other instrument will they need to measure the speed of sound?
    b) Describe the experiment.

11.2.19) Light moves from a material with a larger refractive index into one with a smaller refractive index.
    a) In this situation, does the angle that light has to the normal increase, decrease or stay the same?
    b) What about if it travels from a material with a smaller refractive index into one with a larger refractive index?
    c) What happens if the refractive indices are the same?

11.2.20) A family is watching a fireworks display 500 m away. They see the bright flash of the firework.
    a) How long does it take before the family hears the sound?
    b) Why does the family hear the sound at a different time to when they see the flash of the explosion?

11.2.21) Sound travels at a rate of approximately 1 km per second in liquids and about 5 km per second in solids. Approximately how long would it take for sound to travel through 5,200 m of the following?
    a) Air
    b) Salt water
    c) Steel railway lines

There are 4 materials each with slightly different properties in the table below. This table will be needed in the next few questions.

	A	B	C	D
Refractive index	2.1	1.9	1.8	3.9
Colour	Deep Red	Dark Purple	Bright Orange	Light Green
Density /kgm^{-3}	900	1700	1400	1700
Speed of sound/ms^{-1}	600	1100	2500	900

11.2.22) In which material does light travel:
  a) The fastest
  b) The slowest

11.2.23) In which material does sound travel:
  a) The fastest
  b) The slowest

11.2.24) A beam of light is sent through each of the materials and the time taken is recorded. Identify each of the following materials.
  a) 1 m of material, time taken = 1.3×10^{-8} seconds
  b) 10 cm of material, time taken = 6×10^{-10} seconds

11.2.25) In another experiment sound is sent through a sample of each of the materials. Identify each of the materials by their results.
  a) 15 cm, time taken = 6×10^{-5} seconds
  b) 20 cm, time taken = 1.8×10^{-4} seconds
  c) 35 cm, time taken = 3.8×10^{-4} seconds
  d) 200 cm, time taken = 5.8×10^{-4} seconds

11.2.26) An image is created using a single thin converging lens. The image is one half of the size of the object and it is inverted.
  a) Describe the placement of the object in relation to the lens.
  b) Use this information to decide if the image is real or not.

11.2.27) Complete the diagram below and describe the image created.

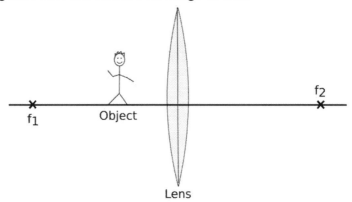

11.2.28) Complete the diagram below and describe the image created.

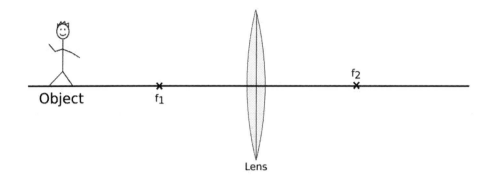

12. More Complicated Systems

Nutcrackers as Force Multipliers

A mechanical force multiplier: In this case, a pair of nutcrackers.

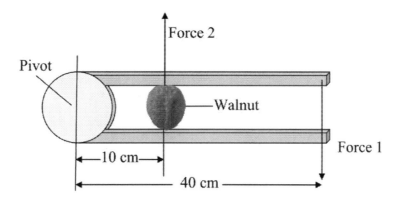

A force is applied to the end of the handles. Here it is 40 cm away from the pivot. In order to stop the handles from getting closer together, the moment because of force 2 (the walnut) must be equal to the moment created by force 1 (your hand squeezing) and in the opposite direction. Here you apply a force of 10 N to the handles (force 1).

moment 2 = moment 1

force 1 x distance 1 = force 2 x distance 2

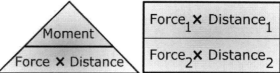

If force 1 equals 10 N, then force 2 from the nut can be calculated.

force 1 x distance 1 = force 2 x distance 2

$$\text{force 2} = \frac{\text{force 1 x distance 1}}{\text{distance 2}}$$

$$= \frac{10 \text{ N x } 0.4 \text{ m}}{0.1 \text{ m}}$$

$$= 40 \text{ N}$$

To stop the handles from moving, the walnut has to create a force of 40 N. You only applied 10 N of force to the handles. The force from the handle has been multiplied to a force that is four times as large.

A Hydraulic Force Multiplier

This is another type of force multiplier which uses liquid to change the force.

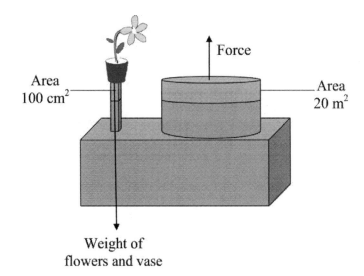

Using the information from the image above, calculate the force that is created on the right hand side of the hydraulic force multiplier by the force due to the weight of the flowers on the left hand side of the hydraulic force multiplier.

Remember that our equation for pressure is pressure = force / area.

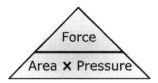

Rearranging the equation gives force = area x pressure

In the picture the pressure is the same at equal heights in the liquid. This is because the liquid is very good at transmitting pressure as it doesn't compress easily and it transmits force. This means, as the two pistons are at equal heights, the pressure underneath them is the same.

To calculate the amount by which the force is multiplied it is approached in this way:

Step 1: Calculate the pressure on the left.
Step 2: Calculate the force on the right.

area is measured in m², there are 10,000 cm² in 1 m²

area on the left = 100 cm²

$$= \frac{100 \text{ cm}^2}{10,000 \text{ cm}^2}$$
$$= 0.01 \text{ m}^2$$

pressure on left $= \dfrac{\text{force}}{\text{area}}$

$= \dfrac{20 \text{ N}}{0.01 \text{ m}^2}$

$= 20{,}000 \text{ Pa}$

force on right = area x pressure
 = 20 m² x 20,000 Pa
 = 400,000 N

One little plant produces enough force to lift an elephant!

This is the same reason that we can use one foot to stop a fast moving car travelling down the highway when we press on the brakes. Our foot acts through a hydraulic force multiplier. The hydraulic brakes only move a small distance compared to your foot but they can clamp down on the brake disk with a large force.

Wire Travelling in a Magnetic Field

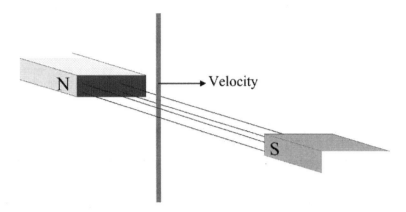

Whenever there is a wire cutting through magnetic field lines there will be a voltage created. The magnetic field lines travel from magnetic North to magnetic South. These lines show the direction that a magnetic compass needle would point if you placed one there.

As a wire moves through a magnetic field it will cut through the magnetic field lines. When this happens the electrons in the wire will experience a force pushing them in one direction. As the electrons are gaining a moving force or a motive force because of this motion, this is called an induced EMF (electromotive force).

The induced EMF is measured as a voltage. So the motion of the wire in the magnetic field creates a voltage and if there is a complete circuit then this voltage will create a current through the wire.

As the voltage is created by the wire travelling through the magnetic field, if the wire was travelling in the opposite direction it would generate a current in the opposite direction.

This is an important thing to know as it is the basis of how electric generators work. So let's examine this.

An Electric Generator

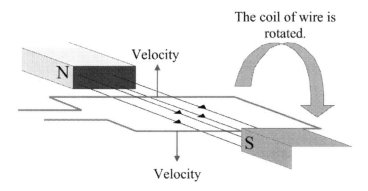

If we manually rotate a coil of wire in a magnetic field, then there will be an alternating voltage induced. When one side travels up, a current is created in one direction around the coil. At the same time the other side of the coil moves down through the wire and creates a current that continues through the loop.

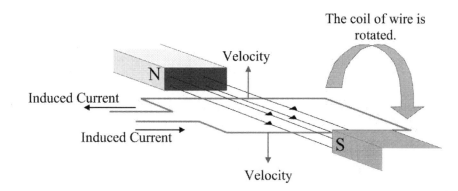

As the coil rotates around, the direction in which it is travelling will change (from up on the left hand side to down on the right hand side) as it spins. This means that as the direction that it is cutting the field changes, or alternates, the value of the voltage will also change, or alternate between a positive and negative value. The positive and negative of the value just describes the direction of force experienced by the electrons in the wire. This will in turn, create an alternating current (AC).

If I increase the number of turns in the wire, or the speed the coil of wire turns, or the strength of the magnetic field it will increase the voltage induced and the current created.

The Magnetic Field of a Current in a Wire

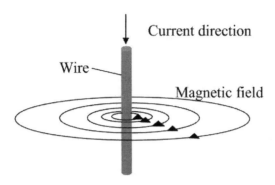

A current in a wire creates a magnetic field. The direction arrows on the magnetic field lines show the direction that a compass would point if it was placed in the field.

A wire, carrying a current, will feel a force when it is placed in between 2 magnets because it has its own magnetic field. The magnetic field from the wire interacts with the magnetic field from the magnets.

In the picture below, the round circle with a dot at the centre represents a wire with a current travelling out of the book. The round circle with a cross in it represents a wire carrying a current moving into the book.

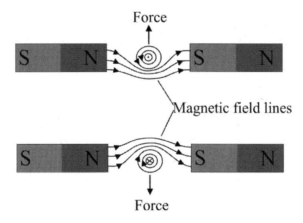

In both cases the wire experiences a force. When the current is travelling in a different direction, the wire feels a force in the opposite direction.

This is an important thing to know as it is the basis of how electric motors work.

An Electric Motor

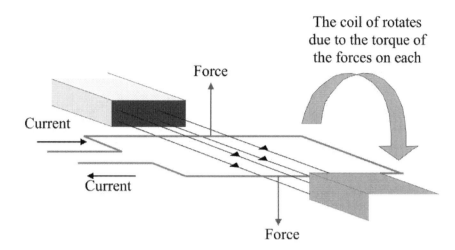

A current in a wire in a magnetic field will generate a force which will attempt to move it.

A current travelling through a coil in a magnetic field will experience a force upwards on the one side and a down force on the other. The effect of this is that the coil of wire will begin to rotate.

Once it reaches the top (the coil is side on to the magnetic field) the current must change direction or it will stop moving.

This is done either using an AC voltage across the motor which creates an alternating current which will change direction anyway, or a split ring commutator which manually changes the direction of the current (switches the wire contacts).

Electromagnets

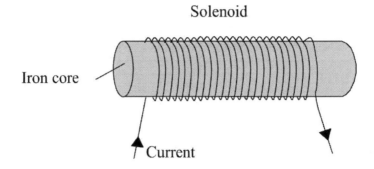

A current moving in the wire will create a magnetic field because as the current in the solenoid travels in the same direction around the solenoid, it makes the magnetic field stronger. It is like having many different wires all with a current going in the same direction so the magnetic field gets stronger.

This also makes the magnetic field the same shape as a bar magnet.

The iron core of the solenoid also makes the magnetic field stronger as it adopts the same magnetic field as the wire while there is a current. This is because the field from the wire forces the magnetic domains inside the iron to line up. This means that the solenoid with a current behaves as a big magnet. It has a magnetic field just like a normal bar magnet with a north pole at one end and a south pole at the other.

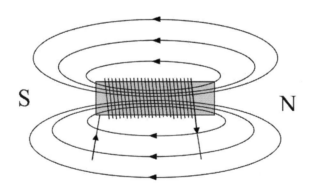

When the current is switched off, the external magnetic field goes away and the iron loses its magnetism (it is a soft magnet and does not hold magnetism). The electromagnet is no longer magnetic if there is no electric current.

A Transformer

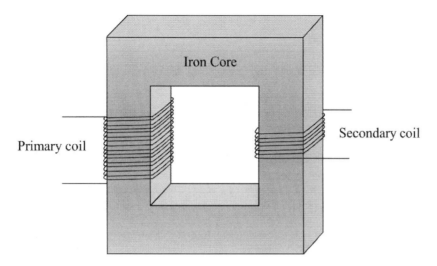

An alternating current travelling in the primary coil creates an alternating magnetic field (the north and south pole sides keep switching with the direction of the current). The magnetic field is trapped inside the iron core. There is a constantly changing magnetic field contained inside the primary coil that is transferred, inside the iron core, to the secondary coil.

Thus the constantly changing magnetic field inside the primary coil creates a constantly changing magnetic field within the secondary coil. This alternating magnetic field will induce an alternating voltage and therefore an alternating current in the secondary coil.

The amount of voltage in the secondary coil depends on the number of turns on the secondary coil compared to the number of turns on the primary coil. If there are more turns on the secondary coil there will be more voltage (but less current) created in the secondary coil than is supplied to the primary coil. This is a step up transformer.

If there are fewer turns on the secondary coil, then the voltage created will be smaller (but the current larger) than the supply voltage. This is a step down transformer.

This will only work with an alternating voltage as this creates an alternating magnetic field. It will not work with DC electricity.

Questions: More Complicated Systems

12.1) Explain how an electromagnet works.

12.2) Explain how a motor works.

12.3) Explain how a generator works.

12.4) Explain how a pair of nutcrackers work.

12.5) Explain how an electrical transformer works.

12.6) A force multiplier is made from a liquid and 2 pistons, one large piston and one small piston both at the same height. A force is applied to the small pistons and a larger force is created at the large piston. Explain how it works.

Answers: More Complicated Systems

12.1) Explain how an electromagnet works.

Something happens:
The current in the wire generates a magnetic field around the wire.

Which has an effect:
As the wire is coiled into a solenoid the magnetic field created is the same as a bar magnet; with a north and a south pole. The same magnetic field is created in the iron core as the magnetic domains line up with the magnetic field of the solenoid.

Which means:
This means that the electromagnet will be magnetic as long as the current is turned on.

Therefore:
When the current is turned off the electromagnet will no longer be magnetic.

Final answer:
The current in the wire generates a magnetic field around the wire. As the wire is coiled into a solenoid the magnetic field created is the same as a bar magnet; with a North and a South pole. The same magnetic field is created in the iron core as the magnetic domains line up with the magnetic field of the solenoid.

This means that the electromagnet will be magnetic as long as the current is turned on. When the current is turned off the electromagnet will no longer be magnetic.

12.2) Explain how a motor works.

Something happens:
A current passes through the coil of wire which is inside a magnetic field.

Which has an effect:
This creates a magnetic field around the coil of wire which interacts with the magnetic field the coil is in.

Which means:
The coil of wire experiences an upward force on one side of the coil and a downward force on the other side.

Therefore:
The coil experiences a turning effect and begins to turn. When the coil is at 90 degrees to the magnetic field lines the current changes direction so that the coil continues to turn.

Final Answer:
A current passes through the coil of wire which is inside a magnetic field. This creates a magnetic field around the coil of wire which interacts with the magnetic field the coil is in. The coil of wire experiences an upward force on one side of the coil and a downward force on the other side. The coil experiences a turning effect and begins to turn. When the coil is at 90 degrees to the magnetic field lines the current changes direction so that the coil continues to turn.

12.3) Explain how a generator works.

Something happens:
As the coil of wire is rotated it cuts through the magnetic field lines.

Which has an effect:
The electrons in the coil experience a force.

Which means:
This creates a voltage and therefore a current in the coil.

Therefore:
The turning coil of the generator in the magnetic field generates electricity.

Final Answer:
As the coil of wire is rotated it cuts through the magnetic field lines. The electrons in the coil experience a force. This creates a voltage and therefore a current in the coil. The turning coil of the generator in the magnetic field generates electricity.

12.4) Explain how a pair of nutcrackers work.

Something happens:
A force is applied to the end of the nutcrackers. This creates a moment or turning force that causes the nutcrackers to close.

Which has an effect:
To stop the nutcrackers closing, the nut must create an equal moment in the opposite direction.

Which means:
(moment = force x distance) The nut is much closer to the pivot so the distance is smaller and the force created by the nut must be much larger.

Therefore:
The nut experiences a much larger force than the original force applied to the handles. The force has been multiplied.

Final Answer:
A force is applied to the end of the nutcrackers. This creates a moment or turning force that causes the nutcrackers to close. To stop the nutcrackers closing, the nut must create an equal moment in the opposite direction. (moment = force x distance) The nut is much closer to the pivot so the distance is smaller and the force created by the nut must be much larger. The nut experiences a much larger force than the original force applied to the handles. The force has been multiplied.

12.5) Explain how an electrical transformer works.

Something happens:
There is alternating electrical current in the primary coil.

Which has an effect:
This creates an alternating magnetic field around the coil which is trapped in the iron core of the transformer.

Which means:
The alternating magnetic field of the primary coil is transferred to the secondary coil by the iron core. This means that the secondary coil now contains an alternating magnetic field.

Therefore:
The electrons in the secondary coil experience a force due to the alternating magnetic field which creates an alternating voltage in the secondary coil.

Final Answer:
There is alternating electrical current in the primary coil. This creates an alternating magnetic field around the coil which is trapped in the iron core of the transformer. The alternating magnetic field of the primary coil is transferred to the secondary coil by the iron core. This means that the secondary coil now contains an alternating magnetic field. The electrons in the secondary coil experience a force due to the alternating magnetic field which creates an alternating voltage in the secondary coil.

12.6) A force multiplier is made from a liquid and 2 pistons, one large piston and one small piston both at the same height. A force is applied to the small piston and a larger force is created at the large piston. Explain how it works.

Something happens:
A force is applied to the small piston.

Which has an effect:
As pressure = force / area, increasing force on the piston creates a pressure in the liquid. The pressure in the liquid is transmitted through the liquid and acts in all directions.

Which means:
There is a resultant pressure created on the larger piston.

Therefore:
As force = area x pressure and the area of the second piston is larger, there is a larger force created at the second piston. The force has been multiplied with a hydraulic force multiplier.

Final Answer:
A force is applied to the small piston. As pressure = force / area, increasing force on the piston creates a pressure in the liquid. The pressure in the liquid is transmitted through the liquid and acts in all directions. There is a resultant pressure created on the larger piston. As force = area x pressure and the area of the second piston is larger, there is a larger force created at the second piston. The force has been multiplied with a hydraulic force multiplier.

12. Bonus Material

One Memory Aid to Rule Them All

The following is a story. It will help you to remember the mnemonics (memory aids) that you have learned using this book. Take time to remember each step. It is based on a method of memorization first developed in ancient Greece over 2,000 years ago.

The method works by getting you to visualize your environment and attach memory cues to your own location and experience. As you read through the following story, practice visualizing what is happening and the order it is happening in.

As you read through the following section try to visualize as clearly as possible the events that are described. Imagine hearing the noises as you open a door, imagine the voices of people talking. Imagine the smells and sounds around you and feel the items you touch.

Each of the steps will lead back to one of the ways you have now learned to remember the equations.

The story:

You get out of bed in your pyjamas. You have a yawn and go to the bathroom to brush your teeth and have a shower. You <u>u</u>se <u>e</u>lectric <u>t</u>oothpaste to get them nice and clean and then when you are finished you open the shower door.

<u>Use</u> <u>e</u>lectric <u>t</u>oothpaste.

And there you see Colin who is in the shower, fully clothed, with an MP3 player. He changes the tune on his mp3 player and starts dancing while he turns the shower on. You slap yourself on the head. You should have known because <u>e</u>very <u>m</u>orning <u>C</u>olin <u>changes</u> tune in the shower. <u>W</u>e <u>f</u>ind <u>d</u>ancing <u>w</u>ill <u>t</u>ake <u>p</u>lace.

<u>E</u>very <u>m</u>orning <u>C</u>olin <u>changes</u> <u>t</u>une.
<u>W</u>e <u>f</u>ind <u>d</u>ancing
<u>w</u>ill <u>t</u>ake <u>p</u>lace.

The water is making his clothes all wet and it is falling all over his <u>head</u>, his <u>neck</u> and his <u>legs</u> but his mp3 player is waterproof so there is nothing to worry about.

Colin has <u>legs</u>, a <u>neck</u> and a <u>h</u>ead.

You close the shower curtain and let Colin finish. With the sound of humming in the background you go to the kitchen to get some food. When you get to the kitchen, a mouse is cooking at the stove and watching the mouse is a kangaroo and a penguin. As we know, <u>e</u>very <u>m</u>ouse <u>c</u>an <u>c</u>ook. The penguin and kangaroo look excited and you can see them drooling. It is a good <u>p</u>enguin so it will <u>e</u>at <u>h</u>ot <u>m</u>ackerel and gravy and the <u>k</u>angaroo will <u>have</u> <u>m</u>eat and <u>v</u>egetables <u>too</u>. It smells terrible.

<u>E</u>very <u>m</u>ouse <u>c</u>an <u>c</u>ook.
<u>G</u>ood <u>p</u>enguins <u>e</u>at <u>h</u>ot <u>m</u>ackerel and <u>g</u>ravy.
<u>K</u>angaroos **<u>have</u> <u>m</u>**eat and <u>v</u>egetables **<u>too</u>**.

You open the fridge to get some food but all you see is another penguin. This one is beating a sheet of aluminium with a hammer and there is a lump of lead on the floor. <u>A</u> penguin <u>b</u>eating <u>aluminium</u> <u>lead</u> <u>by</u> <u>chance</u>? How unusual! It looks at you and indicates that it wants you to close the door to stop the heat getting in so you close the door and turn around.

<u>A</u> penguin
<u>b</u>eating **<u>aluminium</u>**
gets **<u>lead</u>**
<u>by</u> <u>chance</u>.

You see your friend Mr Quark sitting at the table. He is covered in pudding. He shouts loudly, 'I love my puudddin!'. You tell him to be a little quieter as it is first thing in the morning but you know how much all <u>quarks</u> love their <u>puuddin</u>. Mr Quark apologises but continues to make a mess with his morning pudding.

Quarks love pudding!

Next to Mr Quark is a strange creature that looks like a tent with a head. It's jumping on a broom singing a little song, 'the <u>quark</u> was <u>strangely</u> charming from the <u>bottom</u> to the <u>top</u>. It jumped <u>up</u>, it jumped <u>down</u> and it jumped all around, while the <u>mutent</u> jumped on the mop'. You ask the creature what it has been <u>up</u> <u>to</u> but it just keeps jumping and singing.

The **<u>quark</u>** was **<u>strangely</u> <u>charming</u>**,
From the **<u>bottom</u>** to the **<u>top.</u>**
It jumped **<u>up</u>**, it jumped **<u>down</u>**,
and it jumped all around,
while the **<u>mu</u>-<u>ten</u>**t jumped on the mop**.**

What have you been **<u>up</u> <u>two</u>**?

You are getting hungry so you open the cupboard. There are only two things in the cupboard, some canned vegetables and some pickled carrots. You reach for the pickled carrots because they taste nicer. Everyone knows that you <u>c</u>an't <u>t</u>aste <u>c</u>arrots if you <u>e</u>at <u>c</u>anned <u>v</u>egetables. You should <u>e</u>at <u>t</u>hem <u>p</u>ickled. You open the jar, eat one carrot and put the rest back.

<u>C</u>an't <u>t</u>aste <u>c</u>arrots
<u>E</u>ating <u>c</u>anned <u>v</u>egetables.
<u>E</u>at <u>t</u>hem <u>p</u>ickled.

You look down and a group of different types of rabbits and bunnies sitting patiently staring at you. They really like carrots and you put them back! There are <u>v</u>ery <u>i</u>mportant <u>r</u>abbits, <u>v</u>ery <u>large</u> rabbits and also <u>small</u> rabbits, and they are all watching you.

<u>V</u>ery <u>i</u>mportant <u>r</u>abbits
<u>V</u>ery **<u>large</u>** rabbits
<u>A</u>lso **<u>small</u>** rabbits

You sigh and take the jar back out, open it and give each bunny a carrot. They look very pleased and hop away. The last bunny gives you a gift of thanks. It's an orange rabbit and it is very soft and fluffy. You know very interesting bunnies give you orange rabbits. The bunny gives you a little nose twitch and hops away to enjoy his pickled carrot.

<u>V</u>ery <u>i</u>nterested <u>b</u>unnies <u>g</u>ive <u>y</u>ou <u>o</u>range <u>r</u>abbits.

You go back to your room and to get ready for the day. You put on your favourite aluminium belt and as you are putting on your socks, nine little green aliens come in through the window and bounce up and down on your bed while they happily chant your name. It is as if these nine green men know nothing except maybe my name. You direct them back through the window and lock it.

<u>9</u> <u>g</u>reen <u>m</u>en <u>k</u>now <u>no</u>thing ex<u>c</u>ept <u>m</u>aybe <u>m</u>y <u>n</u>ame.

As you go to leave the house you see your friend. They need to practice dancing. They say they need to find a partner to practice with because partners have a great dance pivoting in and pivoting out.

<u>F</u>ind <u>a</u> <u>p</u>artner.
<u>P</u>artners <u>h</u>ave a <u>g</u>reat <u>d</u>ance
<u>p</u>ivoting <u>in</u>, <u>p</u>ivoting <u>out</u>.

You take them into the front room which is full of Vikings visiting your family. You put on some fast music to make those Vikings move and make the Vikings dance. You must find dancers for every song. You give your friend the job of finding dancers and you leave your friend with the Vikings.

<u>M</u>ake <u>V</u>ikings <u>m</u>ove.
<u>M</u>ake <u>V</u>ikings <u>d</u>ance.
<u>M</u>ust <u>f</u>ind <u>d</u>ancers
<u>f</u>or <u>e</u>very <u>s</u>ong.

You leave the house and you start to walk down the street. You are singing your favourite song. As you walk you pass a man who has my guitar. You remember that you let him borrow it a while ago. You realise that just by singing a tune down the street you found a man with my guitar!

<u>S</u>ing <u>a</u> <u>t</u>une
<u>d</u>own <u>t</u>he <u>s</u>treet.
<u>F</u>ind <u>a</u> <u>m</u>an
<u>w</u>ith <u>m</u>y <u>g</u>uitar.

Next to the man is a big carousel all lit up with flashing lights. How exciting! You will have to come back later to have a ride but you are busy right now. You remember that your favourite speed on the carousel is fast.

Remember your <u>fav</u>ourite <u>speed</u>.

You keep walking down the street to a corner. On the corner are two workmen with enormous hats. You hear them complaining loudly. "Silly new rules" one of them exclaims. The other nods enthusiastically and says some of the fed up workers never signed in and never signed out. They continue their grumbling and you move on.

<u>S</u>illy <u>new</u> <u>r</u>ules.
<u>S</u>ome <u>f</u>ed up <u>w</u>orkers
<u>N</u>ever <u>sign</u>ed <u>in</u>, never <u>sign</u>ed <u>out</u>.

Once you get farther from your house you wonder which path you should take to your destination. There are three possible routes; the first is the fastest and the other two take you through interesting focal points.

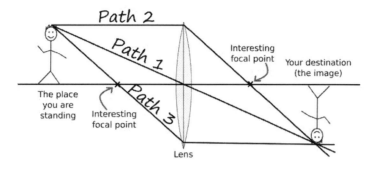

As you wander along the street you see your friend, Short Gavin, sitting in the front garden pretending he is on a throne. He sees you and asks if you are coming over to play. You tell him you are busy and you can't. This makes Short Gavin sad but he pretends he didn't want to play anyway. He says to you, "good, I am glad you are not coming over. I execute unwelcome visitors in a mad rage!"

Short Gavin e**x**ecutes **u**nwelcome **v**isitors **i**n a **m**ad **r**age.

You continue on. You are almost at the store but you see a sign for a new museum you have not visited before. It says, 'view ugly and amazing teapots! See ugly teapots and have a tea too! To all schools and you too, visit today to see us and view teapots'. Teapots are not really your thing though so you decide to pass up on the opportunity.

'**V**iew **u**gly **and a**mazing **t**eapots!
See **u**gly **t**eapots **and have a** **t**ea **too**!
To all schools **and you too**,
visit **to**day
to see **us and v**iew **t**eapots

You keep walking as you are trying to keep up your momentum to buy your favourite food ice cream at the store. You know that momentum is just like your favourite food, it doesn't change.

Total **momentum** is like your favourite food. It just **doesn't change**.

You go into the store and buy the ice cream and as you are leaving you see a huge fish tank at the entrance. It is at least a metre high! At the bottom of the fish tank are a few robots. Swimming past the robots is a large fish with a top hat. It's Sapud the Wonderfish. You wonder what makes him so special. It must be his ability to hold a cane and wear a top hat and vest. This is not a normal for fish.

SAPUD the Wonderfish

You roll up your sleeve, reach into the tank and pull up the robots, leaving them next to the tank. You see very visible rust over all of the robots. You hope they will keep working.

Ve**r**y **v**i**s**ible **r**ust **over** all **r**obo**t**s

You set off again because you have to get home. You have left a house full of animals and Vikings and need to make sure it is still standing. As you leave the store you see a man sitting under an apple tree and as you watch an apple falls and hits him on the head. He seems ok though so you carry on home.

> **Newton's 1st law**: An object a rest will remain at rest and an object in motion will continue with the same motion as long as the forces are balanced.
> **Newton's 2nd law:** If an unbalanced force acts on an object, that object will accelerate in the same direction as the force according to F = ma.
> **Newton's 3rd law:** For every action force there is an equal and opposite reaction force.

When you get home you wonder what distance you walked. You see an old lady passing and ask her. She answers, "do you understand it?" You think this that is a very strange response and wonder if she heard you properly.

If you want to find the **distance travelled,** you have to **underst**and it.

You look in the driveway and are very sad. Your parents are selling your first van and have bought another one. You really do prefer your first van over your second van.

I prefer my **first van over** my **second van**.

As you are standing there remembering all of the wonderful trips you have taken in your first van, a car stops next to you and a nosey vampire sticks his head out the window and asks if you know the way to visit Naples. You tell him that it is a very long way away in Italy and you probably can't get there in that vehicle. He looks down his nose at you, rolls up the window and drives away.

No**s**ey **V**am**p**ires **v**i**s**it **N**a**p**les

You shake your head and turn towards the front door of your house. You are about to open the door when you see a big lead gem outside on the porch. You are confused because you did not know that lead could form gems. As you are pondering this you realize that on your journey you have lost your aluminium belt. How strange. You open the front door wondering where it is and you find two pins and a paper hat inside.

Lead gem 0utside
Lost aluminium belt
Found 2 pins and a **paper hat inside**

You put on the paper hat and wander towards the front room to see if the Vikings are still dancing. It looks like it has become pretty lively in there. There is big, powerful alien mouse acting as a bouncer at the door. It is very top heavy. A little bit like an atomic symbol. It is wearing a t-shirt that says' "It's really interesting in Venice". You ask it what is so interesting about Venice but the music is too loud. It thinks you are asking it why it left Venus. Apparently Venus is too over run with alien mice.

It's **r**eally **i**nteresting
in **V**enice.
(In Venus?)
Venus??? **Too over r**un
(with alien mice.)

The alien mouse doesn't want to let you in but your friend is still there. Your friend nods to you and the mouse lets you in. You go to see your friend. You start dancing and you are happy that you know so many physics equations.

The end…or is it?

That's the story. Each part of it links to a way of remembering the different physics equations. You will be amazed at how much you can remember when you try to use this method. Practice it several times. See and interact with the events as you visualize them in your head. Practice writing down the different equations and memory aids that are linked to the story.

A Final Note

We have now reached the end of the book. The purpose of this book wasn't to teach you physics. You probably have a specific textbook that has all of the information you need. This book was designed to help you remember the many equations and concepts and teach you how you can use them easily. Once you know this you should be able to answer pretty much any question your physics teacher throws at you! Physics becomes much easier when you have the right tools and approaches. Your class work, homework and tests should all become much easier and remember that this success is yours because of your hard work.

The brain is an interesting organ. If you visualise each mnemonic and story that goes with the equation, you will find that it is easy to remember the equation. The main thing now is practice. Practice as much as you can.

You will find that once you remember the equations and how to rearrange them, everything else becomes a little bit easier and more fun. Your teacher may wonder what the little scribble about green men knowing your name is at the side of your test paper but they will be impressed with your abilities to answer the questions correctly. For your tests, read through all of the questions quickly and write little helpful notes for when you come back to them (perhaps the equations you need for example), then go through the test and <u>do all of the easy questions first</u>. Once you have done that, complete the questions that are a little tricky and <u>save the difficult questions until the end</u>. This last little hint should make your exams much easier.

I hope that you will enjoy physics as much as I do and that in some small way this book will help.

Good luck!

Michael Reid.

All the Equations in One Place

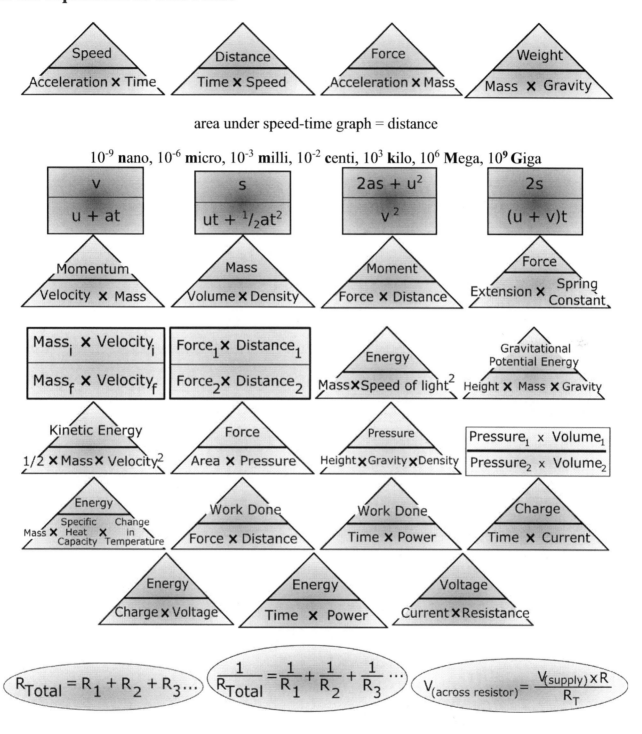

area under speed-time graph = distance

10^{-9} **nano**, 10^{-6} **micro**, 10^{-3} **milli**, 10^{-2} **centi**, 10^{3} **kilo**, 10^{6} **Mega**, 10^{9} **Giga**

Ammeters have small resistance   Voltmeters have large resistance

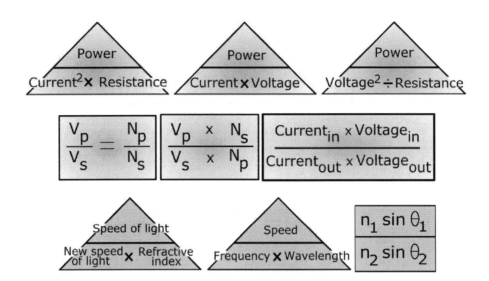

Elastic Potential (strain) Energy Electrical Energy Sound Energy
Gravitational Potential Energy (GPE) Nuclear Energy Heat Energy
Chemical Potential Energy Light Energy Kinetic Energy (KE)

Radioactivity:

Small number: alpha → -2, beta → +1, gamma → 0
Large number: alpha → -4, beta → 0, gamma → 0

Outside the body: (most dangerous) gamma, beta, alpha (least dangerous)
Inside the body: (most dangerous) alpha, beta, gamma (least dangerous)

Alpha radiation → stopped by paper

Beta radiation → stopped by aluminium

Gamma radiation → stopped by lead

Protons: Up, Up, Down   Up Quarks: $+\frac{2}{3}$

Neutrons: Down, Down Up Down Quarks: $-\frac{1}{3}$

Newton's 1st law: An object at rest will remain at rest and an object in motion will continue with the same motion as long as the forces are balanced.

Newton's 2nd law: If an unbalanced force acts on an object, that object will accelerate in the same direction as the force according to F = ma.

Newton's 3rd law: For every action force there is an equal and opposite reaction force.

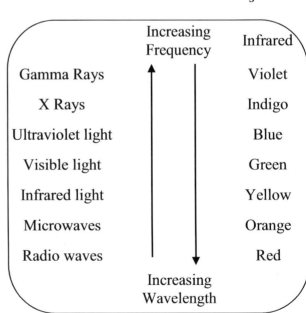

Calculating Area and Volume

2 Dimensional Shapes

Circle

    area = pi x radius^2
    area = ¼ x pi x diameter^2

    pi = 3.142 (approximately…)

Square

    area = length x height

Rectangle

    area = length x height

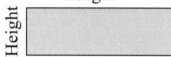

Triangle

    area = ½ x base x height

 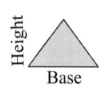

3 Dimensional shapes

Sphere

    volume = 4/3 x pi x radius^3

Cube

    volume = length x base x height

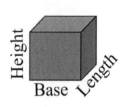

Rectangular box

    volume = length x base x height

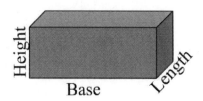

Answers to Bonus Questions

1.1: Speed, Distance, Force and Weight

1.1.1) a) 15 ms^{-1} b) 300 m   c) 3 ms^{-1}
1.1.2) a) 3.57 ms^{-2}
       b) 14 m

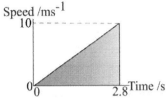

1.1.3) a) 40 ms^{-1} b) Started from rest, no air resistance.
1.1.4) a) 7.5 ms^{-2} b) 10 ms^{-2} or g
1.1.5) a) 19.2 ms^{-1} b) 16 s
       c) 269 m

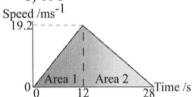

1.1.6) a) 5.98 ms^{-1}
       b) 7.77 m

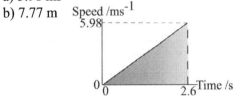

       c) 25.5 m

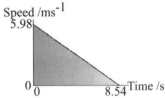

1.1.7) a) 28,000 ms^{-2} b) 10 ms^{-2} or g
1.1.8) a) 2.08 ms^{-2} b) 5 ms^{-2} c) -5 ms^{-1}
1.1.9) 10 ms^{-2} or g
1.1.10) 9,600 N
1.1.11) a) 0.06 ms^{-2} b) 0.1 ms^{-2} c) 1.5 ms^{-2}
1.1.12) 5,220 N
1.1.13) 225 N
1.1.14) a) 0.147 ms^{-2} b) 0.2 ms^{-2} c) 0.5 ms^{-2}
1.1.15) 6 N
1.1.16) 30 kg
1.1.17) a) 3 ms^{-2}
        b) Friction from the sand reduces the force available to accelerate the object.
1.1.18) a) 0.833 ms^{-2} b) 0.25 ms^{-2} c) 0.125 ms^{-2}
1.1.19) a) 10 N   b) 1.62 N   c) 0.42 N
1.1.20) a) 700N   b) 19.2 kN   c) 12.6 GN
1.1.21) 4,620 kg
1.1.22) a) 3 kg   b) 3 kg (Mass does not change with gravity!)

1.1.23) a) 40 N b) 104 N c) 1.1 kN
1.1.24) a) 8.9 ms^{-2} b) 0.89 g c) 0 N
1.1.25) 85.2 kg
1.1.26) 200 N
1.1.27) a) 150 m b) 2,110 m
1.1.28) a) 10.2 ms^{-1} b) 3.31 ms^{-1} faster
1.1.29) Bernard arrives first after a time of 2,778 seconds or 46 minutes 18 seconds.
1.1.30) D
1.1.31) Newton's 1st law: An object at rest will remain at rest and an object in motion will continue with the same motion as long as the forces are balanced.
Newton's 2nd law: If an unbalanced force acts on an object, that object will accelerate in the same direction as the force according to F = ma.
1.1.32) D
1.1.33) B
1.1.34) A
1.1.35) B

1.2: Speed, Distance, Force and Weight
1.2.1) a) 12 ms^{-1} b) 48 m c) 2.4 ms^{-1}
1.2.2) a) 1.25 ms^{-2}
b) 22 m
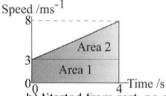
1.2.3) a) 60 ms^{-1} b) Started from rest, no air resistance. c) It will be less due to air resistance.
1.2.4) a) 600 ms^{-2} b) 10 ms^{-2} or g
1.2.5) a) 0.8 ms^{-1}
b) 3.2 m
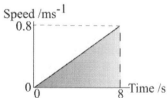
c) 9.6 m
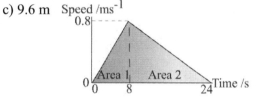
1.2.6) a) 14.4 ms^{-1} b) 16 s c) 180 m
1.2.7) a) 467 ms^{-2}
b) 0.0525 m or 5.25 cm
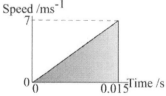
1.2.8) a) 2.45 ms^{-2} b) 1.69 ms^{-2} c) He accelerated faster than he decelerated
1.2.9) 10 ms^{-2} or g
1.2.10) 6,875 N
1.2.11) a) 16,500 N or 16.5 kN b) 75,000 N or 75 kN c) 1,350 N
1.2.12) 2,400 N
1.2.13) 180 N
1.2.14) a) 0.29 ms^{-2} b) 0.8 ms^{-2} c) 6 ms^{-2}
1.2.15) 25 N

1.2.16) The truck has been emptied of its cargo which reduces the total mass of the truck. As force = mass x acceleration, and the force remains constant, the acceleration of the truck increases.

1.2.17) a) 0.83 ms^{-2}
b) Friction from the rough surface reduces the force available to accelerate the cart.

1.2.18) a) 2.07 ms^{-2} b) 0.857 ms^{-2} c) 0.5 ms^{-2}

1.2.19) a) 500 N   b) 1,297.5 N   c) 185 N

1.2.20) a) 15.1 kN   b) 17.7 kN   c) 3.06 x 10^{11} N

1.2.21) a) 175 kg   b) 175 kg   c) 175 kg

1.2.22) 166.7 kg

1.2.23) a) 130 N   b) 18.5 N

1.2.24) 57.8 kg

1.2.25) 680 N

1.2.26) a) 180 m   b) 810 m

1.2.27) a) 9.3 ms^{-1} b) 2.4 ms^{-1} faster

1.2.28) Herbert arrives 1st. It takes 53 seconds until the second fish, Elf, arrives.

1.2.29) A

1.2.30) Newton's 1st law: An object at rest will remain at rest and an object in motion will continue with the same motion as long as the forces are balanced.
Newton's 3rd law: For every action force there is an equal and opposite reaction force.

1.2.31) D

1.2.32) A

1.2.33) D

1.2.34) B

2.1: Equations of Motion and More Complicated Movement

2.1.1) 4.2 ms^{-2}

2.1.2) a) 4 m   b) 12 m   c) 20 m

2.1.3) 5 ms^{-1}

2.1.4) 19.9 ms^{-1}

2.1.5) 0.4 ms^{-1}

2.1.6) -9.78 ms^{-2}

2.1.7) a) -5.58 ms^{-2} b) -14.0 ms^{-2}

2.1.8) 5.8 s

2.1.9) 28 m

2.1.10) 45.5 m

2.1.11) 68 m

2.1.12) 1 ms^{-1}

2.1.13) 25 ms^{-1}

2.1.14) a) 5.17 ms^{-1} b) Acceleration is constant.

2.1.15) 19 ms^{-1}

2.1.16) 5 ms^{-1}

2.1.17) 1.58 ms^{-2}

2.1.18) 0.047 ms^{-2}

2.2: Equations of Motion and More Complicated Movement

2.2.1) 1.87 ms^{-2}

2.2.2) a) 2.5 m   b) 12.5 m   c) 22.5 m

2.2.3) 7.35 ms^{-1}

2.2.4) 17.9 ms^{-1}

2.2.5) 0.7 ms^{-1}

2.2.6) 9.83 ms^{-2}
2.2.7) a) 3.56 ms^{-2} b) 16.1 ms^{-2}
2.2.8) 6.24 s
2.2.9) 4.64 m
2.2.10) 17.9 m
2.2.11) 32.5 m
2.2.12) 2.5 ms^{-1}
2.2.13) 9.29 ms^{-1}
2.2.14) a) 1.65 ms^{-1} b) Acceleration is constant.
2.2.15) 33.5 ms^{-1}
2.2.16) 4 ms^{-1}
2.2.17) 2.7 ms^{-2}
2.2.18) 0.725 ms^{-2}

3.1: Momentum, Force, Moment and Mass
3.1.1) 20 Nm
3.1.2) a) 36 Nm   b) 9 Nm   c) 27 Nm
3.1.3) Yes, it does balance. Both the anticlockwise moment and the clockwise moment are equal to 0.224 Nm
3.1.4) a) 15 Nm   b) 31.3 cm (this is 18.75 cm from the pivot point to 3 s.f.)
3.1.5) a) Yes   b) 29.4 N
3.1.6) 180 kg
3.1.7) 0.23 m^3
3.1.8) 15,400 kgm^{-3}
3.1.9) a) 12,500 kgm^{-3}, 2.5 gcm^{-3} b) 3.75 kg
3.1.10) 0.075 m^3
3.1.11) 1 gcm^{-3}
3.1.12) a) 35.2 kNs   b) 352 kJ   c) 35.2 kNs   d) 176 kJ, this is ½ of the value of the KE of the car.
3.1.13) 1.8 ms^{-1}
3.1.14) 2.88 GNs
3.1.15) 4.29 cms^{-1}
3.1.16) 0.261 ms^{-1}
3.1.17) a) 100 Nm^{-1} b) 10 cm
3.1.18) a) 0.2 m   b) 0.32 m
3.1.19) 0.5 m
3.1.20) a) 7.2 N   b) 720 g
3.1.21) a) 147 Nm^{-1} b) 58.8 N

3.2: Momentum, Force, Moment and Mass
3.2.1) 9 Nm
3.2.2) a) 24.3 Nm   b) 97.2 Nm   c) 72.9 Nm
3.2.3) No, it does not balance. The clockwise moment is 0.728 Nm and the anticlockwise moment is 0.713 Nm. While they are close, they are not the same so the beam does not balance.
3.2.4) a) 4.5 Nm   b) 43.6 cm (this is 6.4 cm from the pivot point)
3.2.5) a) 3 Nm   b) 10 N
3.2.6) 1,224 kg
3.2.7) 1.09 m^3
3.2.8) 1,525 kgm^{-3}, 1.525 gcm^{-3}
3.2.9) a) 8,571 kgm^{-3}, 8.57 gcm^{-3} b) 2,571 g or 2.57 kg
3.2.10) 0.0167 m^3
3.2.11) 13.69 gcm^{-3}

3.2.12) a) 56,400 kgms^{-1} or 56,400 Ns   b) -39,900 Ns   c) 3.17 ms^{-1}
3.2.13) 2.17 ms^{-1}
3.2.14) 8.89 x 10^5 Ns or 889 kNs
3.2.15) 0.023 ms^{-1}
3.2.16) 2.5 ms^{-1}
3.2.17) a) 40 Nm^{-1} b) 3.75 cm
3.2.18) a) 23 cm   b) 25.6 cm
3.2.19) 37.5 cm
3.2.20) a) 3.6 N   b) 360 beads
3.2.21) a) 100 Nm^{-1} b) 45 N

4.1: Kinetic Energy, Potential Energy and Energy Stored in Mass

4.1.1) 80 J
4.1.2) a) 48 kJ
   b) Energy is stored as Chemical Potential Energy in the muscles. This is where the energy came from to allow the worker to climb the stairs.
4.1.3) a) 0 J   b) 160 J   c) 20 kJ
4.1.4) a) 2.4 kJ
   b) GPE → KE as the object falls → Heat and sound when the object lands on the ground.
4.1.5) a) 141 ms^{-1} b) 283 ms^{-1}
   c) No, air resistance will reduce the speed and the ball will reach a terminal velocity.
4.1.6) a) 34.6 ms^{-1} b) 34.6 ms^{-1}, same
4.1.7) a) 69.4 J   b) reduce after it has been fired due to air resistance
4.1.8) a) 5,450 J   b) 196 kJ   c) 1.02 x 10^{10} J or 10.2 GJ
4.1.9) a) The fighter jet has the most kinetic energy at 1.03 GJ
4.1.10) 24.5 ms^{-1}
4.1.11) a) 10 ms^{-1} b) 3,000 J   c) 750 J   d) ¼ of the value
4.1.12) a) 400 kJ   b) 250 kJ, car has the most energy.
4.1.13) a) 1 J   b) 4 J   c) 16 J
4.1.14) a) 1 g   b) 9 x 10^{13} J   c) 180 km
4.1.15) a) 5,560 kg   b) 5.56 x 10^6 kg
4.1.16) a) 1 - Energy from the sun won't run out (any time soon!)
      2 - Supply of energy from the sun is free.
      3 – No greenhouse gases produced.
   b) 1 - Solar panels are expensive
      2 - Only work during the day
4.1.17) Elastic Potential Energy → KE → GPE → KE → Heat and sound as it lands
4.1.18) a) Chemical potential energy → light and heat.   b) Infrared radiation.
4.1.19) Electrical energy → Heat and light, light is useful, heat is not (not for a light bulb anyway!)
4.1.20) Infrared radiation, heat lamp.
4.1.21) KE train → Heat and sound

4.2: Kinetic Energy, Potential Energy and Energy Stored in Mass

4.2.1) 310 J
4.2.2) a) 270 kJ
   b) Energy is stored as chemical potential energy in the muscles. This energy allows a person to climb the CN Tower.
   c) 2.41 GJ
4.2.3) a) 50 kJ   b) 3 J   c) 180 kJ

4.2.4) a) 2.1 kJ
b) GPE → KE as the object falls → Heat and sound when the object lands on the ground.
4.2.5) a) 469 ms^{-1}
b) No. Air resistance would rapidly reduce the speed and kinetic energy of the ice. (The ice would quickly reach terminal velocity.)
4.2.6) a) 20 ms^{-1} b) 20 ms^{-1}, (GPE = KE so mass is not important)
4.2.7) a) 0.69 J   b) Kinetic energy will be reduced as energy is lost due to air resistance
4.2.8) a) 315 J   b) 294 kJ   c) 50 kJ
4.2.9) a) The ball has the most KE with 20 J. (The can has 2 J and the weight has only 0.2 J)
4.2.10) 14.1 ms^{-1}
4.2.11) a) 1.25 ms^{-1} b) 46.9 J
c) Frank has slightly higher average KE as the dog walker has an average KE of 36.3 J.
4.2.12) a) 18 MJ (or 18,000,000 J)
4.2.13) a) 90 J   b) 250 J   c) 640 J
4.2.14) a) 2.7 x 10^{12} J or 2,700 GJ   b) 19,286 s or 5 hr 21 min 26 s
4.2.15) 4.4 x 10^9 kg
4.2.16) a) 1 - Energy from the wind won't run out
2 - Supply of energy from the wind is free.
3 - No greenhouse gases produced
b) 1 - Wind power is dependent on the weather
2 - Need large amounts of land
3 - Can affect wildlife
4 - Noisy
5 - Expensive to build
6 - Some people see them as ugly
4.2.17) GPE → KE + Sound + Heat as it lands
4.2.18) a) Chemical potential energy is converted to heat energy and then to kinetic energy in steam when coal is burned. This turns a turbine, which turns a generator, which produces electricity.
b) Chemical potential energy → heat energy → KE in steam → KE in turbine → KE in generator → electrical energy
4.2.19) Electrical energy → light (useful) + sound (useful) + heat (not useful)
4.2.20) Chemical potential energy stored in muscles → KE as you are climbing → GPE when you increase your height
4.2.21) KE of car → Heat and sound

5.1: Pressure in Solids, Liquids and Gas
5.1.1) 12.5 kPa
5.1.2) 531 Pa
5.1.3) 21.2 kPa
5.1.4) 93.8 kPa
5.1.5) 2.5 MPa
5.1.6) 100 m^2
5.1.7) a) 720 N   b) 72 kg
5.1.8) 7.5 kPa
5.1.9) 22.5 kN
5.1.10) I can lift a mass of up to 22.5 kN/10 ms^{-2} = 2,250 kg as long as I use the full area of the mattress.
a) Yes   b) No   c) No
d) Must be flat under the vehicle so that there is contact with the complete surface. In practice this may be tricky as the underside of the car is not totally flat.
5.1.11) 25 m

5.1.12) 800 kgm^{-3}
5.1.13) 158 kPa
5.1.14) a) 125 kPa   b) 120 kPa   c) 106 kPa – Minimum   d) 640 kPa – Maximum
5.1.15) a) 102 kPa   b) 6 kPa   c) 30 kPa
5.1.16) a) 6 m^2 b) 24.6 kPa   c) 24.6 kN   d) 32.8 kPa   e) 32.8 kN   f) 8.2 kN
5.1.17) 1 MPa, 3 MN
5.1.18) a) 150 m^3 b) 30 m^3
5.1.19) a) 1.13 m^3 b) It could burst.
5.1.20) 246 m^3

5.2: Pressure in Solids, Liquids and Gas

5.2.1) 2.08 kPa
5.2.2) 10.6 kPa
5.2.3) a) 2 kPa   b) Pressure increases over time as the plant grows
c) Most of the mass will be from carbon dioxide in the air. The plant also takes in water and other nutrients that are added to the soil.
5.2.4) 5,750 Pa (to 3 S.F.)
5.2.5) 28.6 m^2
5.2.6) 5 cm
5.2.7) a) 1,200 N   b) 120 kg   c) 1,200 N
5.2.8) 125 kPa
5.2.9) 5 MPa
5.2.10) 6 supports are needed
5.2.11) 15 m
5.2.12) 8,000 kgm^{-3}
5.2.13) 13,760 Pa
5.2.14) a) 37.5 kPa   b) 40 kPa   c) 105.8 kPa – This is the highest pressure
d) 76.4 kPa
5.2.15) a) 384 kPa   b) 72 kPa   c) 120 kPa
5.2.16) a) 13.5 m^2 b) 18.3 kPa   c) 41,175 N   d) 36.6 kPa   e) 82,350 N   f) 41,175 N
5.2.17) a) 200,000 Pa   b) 200,000 N
c) The pressure is the same on both sides of the pane of glass so it does not break.
5.2.18) a) 1.11 m^3 b) 2 m^3 c) 0.5 m^3
5.2.19) The balloon rises up. The outside pressure decreases so the volume of the balloon continues to increase. Eventually the balloon will pop and it will return back to earth.
5.2.20) 0.00091 m^3 or 910 cm^3

6.1: Work, Energy and Power

6.1.1) 36 kJ
6.1.2) 450 kJ
6.1.3) 4.76 K
6.1.4) 588 Jkg^{-1}K^{-1}
6.1.5) 439 Jkg^{-1}K^{-1}
6.1.6) 8.59 MJ
6.1.7) 0.128 kg
6.1.8) 12.0 kJ
6.1.9) a) 120 MJ   b) 132 MJ, B does the most work   c) 100 MJ
6.1.10) a) 305 kJ
b) No, 800 kJ of energy provided by the ice cream is substantially more than the 305 kJ of energy required to climb the Empire State building.

*An interesting side note here however. The human body is actually about 35% efficient (we generate a lot of heat!) If we factor this into the equation, then we would only get about 280 kJ from the ice cream to use to climb and we can combine climbing the Empire State building with eating 100 g of ice cream as a way to not gain weight. Sadly, however, most tubs of ice cream are substantially larger than 100 g! (We're going to need a bigger building).

6.1.11) You are holding the book stationary so if we blindly apply the equation for work done we will get WD = 0.4 kg x 10 ms^{-2} x 0 m which will give a work done of 0 J. So if we are doing no work then why would our muscles get tired? This is an interesting question and the answer relates to the way our muscles work. The muscle fibres are not stationary. They are actually contracting and relaxing on a microscopic level. They *are* actually doing work. You will see the book moving slightly up and down or wobbling because of this. So work is actually being done by the muscles.

6.1.12) a) 36 kJ   b) 48 kJ   c) 31.2 kJ
6.1.13) 90 m
6.1.14) 62.5 N
6.1.15) 8.4 kJ
6.1.16) 200 W
6.1.17) 150 W
6.1.18) 520 W
6.1.19) 100 s
6.1.20) a) 127.5 J   b) 11.8 push-ups

6.2: Work, Energy and Power

6.2.1) 20,475 J or 20.5 kJ
6.2.2) 250.9 kJ
6.2.3) 23.8 °C or 23.8 K
6.2.4) 465 Jkg^{-1}K^{-1}
6.2.5) a) 428.6 Jkg^{-1}K^{-1} b) The material is likely iron
c) Heat is lost to heating up the surrounding air and container. This is why the number obtained experimentally in class will generally always be lower than the actual specific heat capacity. The mystery material is much more likely to be iron rather than copper because your value was higher than the actual value for copper.
6.2.6) 6.78 MJ
6.2.7) 169 g
6.2.8) 6,900 J
6.2.9) The car does the most work at 8.4 MJ. (The cyclist does 4.5 MJ and the motorbike does 7.44 MJ)
6.2.10) 3.15 kJ
6.2.11) a) No work was done as the wall didn't move.
b) You are pushing against the wall so your muscle fibres are not stationary. They are actually contracting and relaxing on a microscopic level. The muscles *are* actually doing work.
6.2.12) a) 67 kJ   b) 40.2 kJ   c) 26.8 kJ
6.2.13) 0.15 m or 15 cm
6.2.14) 33.3 N
6.2.15) 300 kJ
6.2.16) 125 W
6.2.17) 292 W
6.2.18) a) 544 W   b) 1,630 W (to 3 S.F.)
6.2.19) The electric motor completes the task first in 20 seconds. The kettle takes 2 minutes, 27 seconds (147 seconds) to boil.
6.2.20) a) 180 J   b) 129 W

7.1: Radioactivity

7.1.1) $^{8}_{5}B$

7.1.2) $^{55}_{27}Co$

7.1.3) $^{84}_{39}Y$

7.1.4) Step 1: Set up the equipment without the source present and record the level of the background radiation (BG radiation) for a period of 1 minute.
Step 2: Place the source in front of the Geiger counter. Record the count over a period of 1 minute and subtract the BG radiation number to get the radioactivity of the source.
Step 3: Place a piece of paper between the source and the counter. The radiation count should decrease. Record the count for a period of 1 minute and subtract the BG radiation count.
The number should be less than that in step 2, but still not close to 0. This tells us that it emits alpha radiation and this has been stopped by the paper.
Step 4: Replace the paper with a sheet of metal a few mm thick. Record the count over 1 minute and subtract the BG radiation count. The number should be very close to the number in step 3. This tells us that the source does not release beta radiation which would have been stopped by the metal sheet. The radiation that is still being counted must be gamma radiation. As long as the count is higher than the BG radiation count we can conclude that gamma radiation is also being emitted. This proves the source is releasing alpha and gamma radiation.

7.1.5) Step 1: Set up the equipment without the source present and record the level of the background radiation (BG radiation) for a period of 1 minute.
Step 2: Place the source in front of the Geiger counter. Record the count over a period of 1 minute and subtract the BG radiation number to get the radioactivity of the source.
Step 3: Place a piece of paper between the source and the counter. The radiation count should decrease. Record the count for a period of 1 minute and subtract the BG radiation count.
The number should be less than that in step 2, but still not close to 0. This tells us that it emits alpha radiation and this has been stopped by the paper.
Step 4: Replace the paper with a sheet of metal a few mm thick. Record the count over 1 minute and subtract the BG radiation count. The number should be very close to 0. This tells us that the source also released beta radiation which has been stopped by the metal sheet.

7.1.6) Step 1: Set up the equipment without the source present and record the level of the background radiation (BG radiation) for a period of 1 minute.
Step 2: Place the source in front of the Geiger counter. Record the count over a period of 1 minute and subtract the BG radiation number to get the radioactivity of the source.
Step 3: Place a piece of paper between the source and the counter. The radiation count should not change. Record the count for a period of 1 minute and subtract the BG radiation count.
The number should be approximately the same as the number in step 2. This tells us that the source does not emit alpha radiation.
Step 4: Replace the paper with a sheet of metal a few mm thick. Record the count over 1 minute and subtract the BG radiation count. The number should be less than that in step 2 but not close to 0. This tells us that the source releases beta radiation which has been stopped by the metal sheet. The radiation that is still being counted must be gamma radiation. As long as the count is higher than the BG radiation count we can conclude that gamma radiation is also being emitted. This proves that the source is emitting beta and gamma radiation.

7.1.7) Impossible to tell an exact figure as we have no reading for the background radiation count.

7.1.8) To find the level of expected radiation that we would normally expect in the location.

7.1.9) After 3 half-lives the radioactive count will have reached safe levels. 300 (initial) → 150 → 75 → 37.5 (3 half-lives), this will take a time of 25 hours and 30 minutes.

7.1.10) No, alpha radiation could not have penetrated the wrapping of the photographic plate.

7.1.11)

Number of Half-lives	Radioactive count /Bq
0	3,200,000
1	1,600,000
2	800,000
3	400,000
4	200,000
5	100,000
6	50,000
7	25,000
8	12,500

7.1.12) The radioactivity will have decreased to 50 after 3 half-lives. This would correspond to a time of 30 minutes. So after 23 minutes it will still be possible to use the isotope.

7.1.13) $^{238}_{92}U \longrightarrow\; ^{234}_{90}Th + ^{4}_{2}\alpha$

7.1.14) $^{219}_{85}At^* \longrightarrow\; ^{219}_{85}At + ^{0}_{0}\gamma$

7.1.15) $^{14}_{6}C \longrightarrow\; ^{14}_{7}N + ^{\;0}_{-1}\beta$

7.1.16) $^{219}_{85}At \longrightarrow\; ^{211}_{82}Pb + 2\,^{4}_{2}\alpha + ^{\;0}_{-1}e$

7.1.17) 2 hydrogen nuclei fuse together to create helium and give out energy in the process. This is called nuclear fusion.

7.1.18) In nuclear fusion small nuclei are forced together to create larger nuclei. This releases energy. In nuclear fission large nuclei are broken apart into two or more smaller nuclei which also release energy.

7.1.19) The radiation released will be gamma radiation. This is the safest type of radiation to have inside the body. It is also the least likely to be absorbed by the body and can therefore be monitored by machine from outside the body.

7.1.20) 1) There is a plentiful supply of uranium.
2) There are no greenhouse gas emissions.
3) The nuclear plant creates jobs.

7.1.21) 1) Very difficult to dispose of nuclear waste.
2) Danger of radioactive contamination if anything goes wrong.
3) Radiation exposure can have a large number of damaging side effects (cancer, damage to DNA etc.)

7.2: Radioactivity

7.2.1) $^{89}_{40}Zr$

7.2.2) $^{20}_{11}Na$

7.2.3) $^{51}_{26}Fe$

7.2.4) Alpha radiation is more ionising than the other types of radiation. This means that it can cause cell damage more easily so it is very dangerous when it is inside the body.

7.2.5) Gamma radiation is the most difficult to stop so it is more likely to penetrate into the body than alpha or beta radiation. While alpha radiation is the most ionising, it is stopped by a piece of paper so is unlikely to pass through the thick skin layer.

7.2.6) Step 1: Measure the background radiation. This is done by taking a reading over a time of 1 minute with no radioactive source present.
Step 2: Measure the activity of the radioactive source using a Geiger counter.
Step 3: Place a piece of paper between the source and the Geiger counter. If the activity reduces to the level of the background radiation, then the radioactive emissions are being stopped by the paper. This tells us that the radiation released by the source is alpha radiation because alpha radiation is stopped by a piece of paper. If the activity is higher than the normal background radiation reading then it could either be beta or gamma radiation.
Step 4: Replace the piece of paper with a sheet of aluminium a few mm thick. This will stop the beta radiation. If the activity reduces to that of the background level with the aluminium but not with the paper, then you have just shown that the radiation being released is beta radiation.

7.2.7) 0.317 counts/second

7.2.8) Rocks, space, hospital equipment, nuclear testing

7.2.9) 240,000 years (10 half-lives)

7.2.10) This is a question that asks for your opinion. This means that you need to choose a stance and explain why you chose it.
Arguments for:
- Gives a lot of energy for very little fuel.
- No greenhouse gas emissions.
Arguments against:
- Radioactive materials have to be handled and stored very carefully. If it is not being stored and used properly, airplanes carry hundreds of people at a time that could be affected.
- Shielding is very heavy.
- Airplanes unfortunately crash sometimes and this would result in release of radioactive material into the environment.

7.2.11)

Number of Half-lives	Radioactive count /Bq
0	8,000,000
1	4,000,000
2	2,000,000
3	1,000,000
4	500,000
5	250,000
6	125,000
7	62,500
8	31,250

7.2.12) Gamma radiation is reduced in intensity by a few cm of lead but you cannot be sure that it would be completely stopped regardless of how much shielding is used.

7.2.13) $^{238}_{92}U \rightarrow ^{238}_{93}Np + ^{0}_{-1}e$

7.2.14) $^{228}_{88}\text{Ra} \rightarrow {}^{224}_{86}\text{Rn} + {}^{4}_{2}\alpha + {}^{0}_{0}\gamma$

7.2.15) $^{192}_{81}\text{Tl} \rightarrow {}^{196}_{79}\text{Au} + {}^{4}_{2}\text{He} + {}^{0}_{0}\gamma$

7.2.16) $^{112}_{48}\text{Cd}^* \rightarrow {}^{112}_{48}\text{Cd} + {}^{0}_{0}\gamma$

7.2.17) 1) Less CO_2 emissions per kwh
2) Much higher energy density.

7.2.18) a) nuclear fusion   b) nuclear fission

7.2.19) Iodine is most suitable to act as a tracer as gamma radiation is safest when inside the body.

7.2.20) Of these isotopes, the safest when outside the body is polonium, then carbon. Iodine is the most dangerous as it releases gamma radiation.

7.2.21) 1) If anything goes wrong, such as an earthquake, there is a large danger of radioactive contamination of the environment for a very large area.
2) Nuclear waste is very difficult to dispose of.
3) Exposure to radiation can cause damage to DNA which can cause mutations, cancer and radiation poisoning.
4) Waste remains dangerous for a very large period of time (hundreds of thousands of years).

8.1: Charge, Energy, Efficiency and Electrical Power 1
8.1.1) 2A
8.1.2) 345 V
8.1.3) 0.2 Ω
8.1.4) 180 C
8.1.5) 1 A
8.1.6) 100 s or 1 min 40 s
8.1.7) 16 s
8.1.8) 1.4 kJ
8.1.9) 21.7 C
8.1.10) 0.36 J
8.1.11) 283 V
8.1.12) 25.5 MV
8.1.13) 2.4 GJ
8.1.14) 12 kJ
8.1.15) 111 W
8.1.16) 625 s or 10 min 25 s
8.1.17) 0.395 or 39.5%
8.1.18) In these examples some energy will be lost to friction creating unwanted heat energy.
8.1.19) 1.25 MW
8.1.20) 5.71 s
8.1.21) a) 500 W   b) 556 W   c) 4.31 A

8.2: Charge, Energy, Efficiency and Electrical Power 1
8.2.1) 3A
8.2.2) 90 V
8.2.3) 5 Ω
8.2.4) 240 C

8.2.5) 1 A
8.2.6) 225 s or 3 min, 45 s
8.2.7) 103 s or 1 min 43 s
8.2.8) 570J
8.2.9) 6.67 C
8.2.10) 0.72 J
8.2.11) 15 V
8.2.12) 122 MV
8.2.13) 66 MJ
8.2.14) 240 kJ
8.2.15) 278 W
8.2.16) 1,250 s or 20 min 50 s
8.2.17) 0.44 or 44%
8.2.18) No. Some energy will be lost as it is converted to light, sound and heat energy from the lightning itself.
8.2.19) 0.118 or 11.8%
8.2.20) 8.75 s (remember that work = force x distance or GPE = mgh)
8.2.21) a) 3.33 kW   b) 3.92 kW (the funicular is only 85% efficient)   c) 11.4 A

9.1: Resistors in Series and Parallel, Voltage Across Resistors
9.1.1) 110 Ω
9.1.2) 30 Ω
9.1.3) 5 Ω

9.1.4) 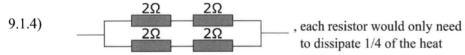, each resistor would only need to dissipate 1/4 of the heat

9.1.5) 1.88 m
9.1.6) 750 Ω
9.1.7) a) 13.6 Ω   b) 7.35 A

9.1.8)

9.1.9) a) The bright one   b) Heat
9.1.10) 17 V
9.1.11) 9 Ω
9.1.12) a) 1 Ω   b) 1 Ω   c) 1 Ω
9.1.13) 2.33 Ω
9.1.14) 6 V
9.1.15) 32.1 V
9.1.16) 11.9 V
9.1.17) a) 6.85 V   b) 5.15 V   c) 6.85 V
9.1.18) a) 2 V   b) 8 V   c) 10 V   d) Yes   e) 2 amps flow through each resistor.
9.1.19) a) 20 V   b) 20 V   c) 20 V   d) same
9.1.20) The fourth option is the correct one. The energy is changing from electrical to light and heat in the bulb.

9.2: Resistors in Series and Parallel, Voltage Across Resistors
9.2.1) 140 Ω
9.2.2) 36 Ω
9.2.3) 12 Ω
9.2.4) The total resistance is equal to each individual resistance.
9.2.5) 1.66 m
9.2.6) 380 Ω
9.2.7) a) 8.33 Ω   b) 4.8 A
9.2.8)

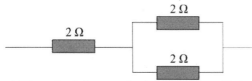

9.2.9) a) The television on the left is more efficient because it produces more light for the energy put in.
b) The extra energy is going towards heating of the wires and other resistors in the television.
9.2.10) 5 V
9.2.11) 20 Ω
9.2.12) a) 0.67 Ω   b) 2.67 Ω   c) 6 Ω
9.2.13) 2.25 Ω
9.2.14) 12 V
9.2.15) 57.1 V
9.2.16) 10.2 V
9.2.17) a) 9.6 V   b) 2.4 V   c) 9.6 V
9.2.18) a) 2.77 V   b) 3.69 V   c) 5.54 V   d) Yes
e) Each resistor has the same current of 0.923 amps. The very slight difference you may get are due to rounding in the calculations.
9.2.19) a) 12 V   b) 12 V   c) 12 V
d) The voltages are the same across parallel parts of the circuit.
9.2.20) B is the correct one. Electrical energy is changed to heat energy but the electrons are not used up.

10.1: Electrical Power 2 and Transformers
10.1.1) The 1 Ω bulb draws 2 A and 4 W.
The 2 Ω bulb draws 2 A and 8 W.
The 3 Ω bulb draws 2 A and 12 W.
10.1.2) 1 Ω
10.1.3) 1.10 A
10.1.4) 3 W
10.1.5) 242 Ω
10.1.6) 133 W
10.1.7) 83.3 W
10.1.8) 200 W
10.1.9) Each lamp will only have ¼ of the voltage that it requires to function fully. This means that they will each only be able to operate at 1/16 of their rated output.
10.1.10) 288 W
10.1.11) 20 V
10.1.12) 12.5 V
10.1.13) 100 W
10.1.14) 350 V
10.1.15) 60 V

10.1.16) $N_p = 115$, $N_s = 12$ Although any with this ratio would be correct.
(i.e. $N_p = 230$, $N_s = 24$ or $N_p = 460$, $N_s = 48$ etc.)
10.1.17) a) The supplied voltage was too high. The current that this created was too large for the bulb to handle.
b) 1,010 V
10.1.18) None! Transformers only operate with alternating current. They do not work with direct current.
10.1.19) a) 400 W   b) Reduce power loss
10.1.20) 4.5 A

10.2: Electrical Power 2 and Transformers

10.2.1) The 4 Ω heater draws 1 A and 4 W.
The 8 Ω heater draws 1 A and 8 W.
The 12 Ω heater draws 1 A and 12 W.
10.2.2) 12 Ω
10.2.3) 2.24 A
10.2.4) 32 W
10.2.5) 80.7 Ω
10.2.6) 25 W
10.2.7) 0.375 W
10.2.8) 510 W
10.2.9) Each lamp will only have part of the voltage that it requires to function fully. This means that they will each only be able to operate at a small amount of their rated output so the light will be dim, the heater won't be warm and the fan will be slow.
10.2.10) 300 W
10.2.11) 11.0 V (to 3 s.f.)
10.2.12) 45 V
10.2.13) 57.6 W
10.2.14) 2,800 V
10.2.15) 375 V
10.2.16) $N_p = 115$, $N_s = 6$ Although any with this ratio would be correct.
(i.e. $N_p = 230$, $N_s = 12$ or $N_p = 460$, $N_s = 24$ etc.)
10.2.17) a) The supplied voltage was too high. The current that this created was too large for the microchip to handle.
b) 4.41 kV
10.2.18) None! Transformers only operate with alternating current. They do not work with direct current.
10.2.19) a) 450 W
b) The thick wires reduce the resistance. This also reduces power loss.
10.2.20) 4.32 A

11.1: Waves, Light and Colours

11.1.1) 2.14×10^8 ms^{-1} b) 1.76×10^8 ms^{-1} c) 3×10^8 ms^{-1}
11.1.2) a) 6.38×10^{14} Hz   b) 5.88×10^{14} Hz   c) 4.62×10^{14} Hz
11.1.3) 4.29×10^7
11.1.4) Violet, yellow, orange, red
11.1.5) Violet, indigo, blue, green
11.1.6) Red, yellow, green, blue
11.1.7) Orange, green, blue, indigo
11.1.8) (Infrared) red, orange, yellow, green, blue, indigo, violet (ultraviolet)
11.1.9) Gamma rays, UV light, blue, red, infrared, radio waves
11.1.10) Gamma rays, X-rays, UV light, infrared, radio waves

11.1.11) There is an electric and a magnetic field both at 90 degrees to each other and to the direction of travel of the wave.
11.1.12) 510 m
11.1.13) 4.44 km
11.1.14) 0.053 s or 53.3 ms
11.1.15) a) 2.26×10^8 ms^{-1} b) 23.5 degrees
11.1.16) 62.7 degrees
11.1.17) a) 1.21   b) 2.48×10^8 ms^{-1}
11.1.18) Set off a firework at a known distance (which is reasonably far away. More than 300 m would be good as a general rule farther is better to reduce the errors from reaction time) and start a stopwatch when it is set off (when the flash is seen). When the sound from the firework is heard, stop the stopwatch. Calculate the speed by using the equation speed = distance/time. This will give you the speed that the sound has travelled. You make the assumption here that the speed of light is so fast that the time the light takes to travel to the observer is tiny enough to be ignored.
11.1.19) a) Angle decreases   b) Angle increases   c) Angle is unchanged
11.1.20) a) 0.588 s   b) Light is much faster than sound
11.1.21) a) 17.7 s   b) 6 s   c) 1.2 s
11.1.22) Under the water there is not enough time delay between the sound reaching the first ear and the sound reaching the second ear. This means that there isn't enough time for the brain to decide the direction of the source of the sound.
11.1.23) a) C   b) D
11.1.24) a) D   b) A
11.1.25) a) B   b) D
11.1.26) a) B   b) D   c) C   d) B
11.1.27) a) The image is between 0 and 1 focal lengths distance from the lens.
b) The image is not real, it is a virtual image, as the lens is being used as a magnifying lens and the image is on the same side as the object.
11.1.28)

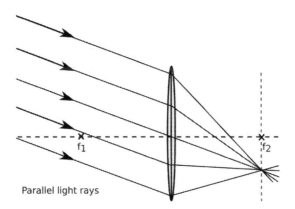

11.1.29)

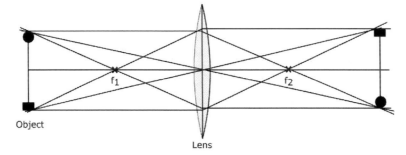

11.2: Waves, Light and Colours

11.2.1) a) 1.88×10^8 ms^{-1} b) 1.70×10^8 ms^{-1} c) 3×10^8 ms^{-1}

11.2.2) a) 4.84×10^{14} Hz   b) 7.5×10^{14} Hz   c) 5.17×10^{14} Hz

11.2.3) a) 2.31×10^8 ms^{-1} b) 0.59×10^8 ms^{-1}

11.2.4) Green, blue, indigo, violet

11.2.5) Violet, yellow, orange, red

11.2.6) Orange, yellow, green, blue

11.2.7) Red, orange, green, blue

11.2.8) (Ultraviolet) violet, indigo, blue, green, yellow, orange, red (infrared)

11.2.9) Gamma rays, X-rays, UV light, yellow light, orange light, radio waves

11.2.10) Gamma rays, X-rays, ultraviolet radiation, visible light, microwaves

11.2.11) It is made of photons which have electric and magnetic fields at 90° to each other and the direction of travel.

11.2.12) 1,020 m

11.2.13) a) at 40°C   b) It takes 0.61 s less time at 40°C than at -1°C

11.2.14) 21.3 ms

11.2.15) a) 1.91×10^8 ms^{-1} b) 22°

11.2.16) 82.3°

11.2.17) a) 1.39   b) 2.16×10^8 ms^{-1}

11.2.18) a) They would need a stopwatch and something to measure distance (metre rule or measuring tape). A calculator is optional.
b) First the students would have to measure a large, known distance away from the building. At least 300 m would be good.
Then the students stand next to each other. One student claps the blocks together and the other times how long it takes to hear the echo.
Finally, the students would calculate the speed of sound using speed = distance x time. The distance here would be twice the distance measured as sound needs to get to the building and back.
They should repeat this procedure at least 3 times.

11.2.19) a) Angle increases   b) Angle decreases   c) Angle is unchanged

11.2.20) a) 1.47 s
b) The speed of light is much faster than the speed of sound.

11.2.21) a) 15.3 s   b) 5.2 s   c) 1.04 s

11.2.22) a) C   b) D

11.2.23) a) C   b) A

11.2.24) a) D   b) C

11.2.25) a) C   b) B   c) D   d) C

11.2.26) a) The object is more than 2 focal lengths away from the lens.
b) The image is inverted so it is a real image.

11.2.27) The image is not inverted. It is virtual. It is larger than the object and it is farther from the lens than the object.

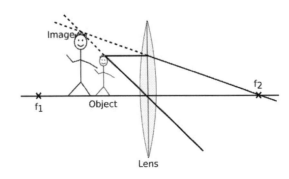

11.2.28) The image is inverted. It is real. It is the same size as the object and it is the same distance away from the lens as the object.

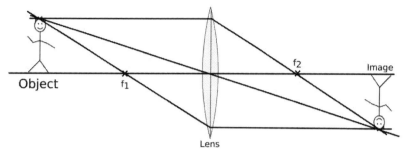

Made in the USA
Middletown, DE
01 February 2019